# "H3" in the Battle Against Old Age

# "H3" in the BATTLE AGAINST OLD AGE

*a dramatic new use for novocain?*

## HENRY MARX

Springer Science+Business Media, LLC

ISBN 978-1-4899-6278-2      ISBN 978-1-4899-6606-3 (eBook)
DOI 10.1007/978-1-4899-6606-3

Copyright © 1960 Springer Science+Business Media New York

Originally published by Plenum Press, Inc. in 1960.

Softcover reprint of the hardcover 1st edition 1960

Library of Congress Catalog Card Number 59-14824

*Old age must be resisted,*
*and its deficiencies supplied.*
— CICERO.

*for my wife,*
*with gratitude*

———

## Acknowledgements

I wish to express my appreciation to the editors of *Lancet,*
*Geriatrics,* and the *British Medical Journal* for permission to
quote from their comments on Professor Anna Aslan's work.
I am also grateful to the many medical practitioners and researchers who have permitted me to quote (sometimes at
considerable length) from their published reports. The photographs on pages 81-97 were provided for publication by Prof.
Anna Aslan and her associates. I can not find adequate words of
appreciation for the cooperation they have extended me.

H. M.

# Contents

**APPENDIX**

## From the Author
### —a statement of purpose

Science has stretched the average life span to nearly seventy years, but as many of those over seventy know so well, merely to exist is not to live. Thus, until medicine can arrest the dissolution of body and spirit that often occurs in the final years of life, until it can make these years worth living, it will have achieved only a pyrrhic victory.

I believe that the work of Professor Anna Aslan and her associates in developing an old age therapy based on regular injections of procaine represents a great stride forward in medicine's march toward an old age period of vitality, usefulness, independence and joy of living for mankind. That is why I wrote about this therapy first in a general magazine in 1958, and why I have written this book.

Since my first article appeared, hundreds of newspaper columns have been filled with stories about the work being done in Bucharest. For the most part, these stories have been either ignorant denunciations, great distortions, half-truths, wishful interpretations, or irresponsible exaggerations. My aim in writing this book was to present the true picture—so that the American public might know how much hope and how much hokum lie behind the sensational headlines on "$H_3$."

I have been privileged to visit the Institute of Geriatrics several times in the past year, and to enjoy the generous cooperation of Prof. Aslan and her associates. They have opened their files to me, patiently answered my innumerable questions. They have permitted me to talk to their patients, and to talk to the same patients when I returned months later. I have also consulted non-Rumanian doctors whose clinical experience with the same therapy in large part corroborates that of Prof. Aslan.

I hope I have succeeded in my attempt to present the material so gathered without prejudice and without emotion.                                    H. M.

# From the Publisher
## —a design for objectivity

In publishing this book on the controversial Anna Aslan therapy for old age diseases, it has been our intention to put the available facts about the "H₃" therapy before the public and before the medical profession, without raising premature or false hopes of the sick and aged and their families. Mr. Marx shares our desire to produce an objective book, and we therefore asked if he would allow his manuscript to be read by a responsible and interested research doctor, with any negative comments on the part of the doctor to be published as an integral part of the manuscript. He wholeheartedly agreed.

A doctor whose specialty is gerontology, and whose interest in treatment of old age diseases has prompted him to investigate the procaine therapy as published by Prof. Aslan, consented to add his comments to this manuscript, his only reservation being that he remain anonymous. He is currently engaged in research in some aspects of aging, and wishes to avoid reader-correspondence on this subject.

Mr. Marx's manuscript was therefore set in type, and the finished galleys were given to this medical scientist so that he could comment where and how he felt it necessary in order to ensure that this would truly be the facts behind the headlines—an objective report on "H₃."

We subsequently received the galleys back, and the following letter, which we have since received his permission to print, explains his position.

---

*From the Medical Monitor*
*—a letter of withdrawal*

I must renege on my promise to serve as a "Medical Monitor" for the book on procaine therapy.

Let me explain.

It sounded like a very good idea when we discussed it: a book by an intelligent medical writer, written for the intelligent layman, and then monitored by a medical scientist. I thought this might be the formula which would enable the American public to evaluate properly the sensational nonmedical reports, and would avoid creation of unrealistic hopes.

I have reviewed the manuscript you have submitted, and find that I have made 97 notations where I would find it necessary to express a reservation, a contradiction, or a challenge. These are by no means limited to points where the author's inadequate knowledge of medicine is involved.

Consider the chapter: "Rejuvenation—an emotionally charged word." It is not surprising that the word induced resistance on the part of medical scientists, or that the press subsequently spoke of such things as "Youth Shots," for Aslan's statement that "novocain reduces the biological age of those treated with it below the chronological age" can be interpreted in only one way: the treatment makes people younger. This is certainly no "modest claim!" Scientific statements bordering on the sensational must be meticulously documented, more so than in the usual situation —to avoid skepticism and disbelief. This entire chapter is a species of involved rationalization.

All this could have been prevented if the word "eutrophic" had been used throughout. It is not an emotionally charged word and implies reversion to normality. It would have been wiser, certainly, to have studiously avoided the word "rejuvenation."

In my opinion, the author has largely failed in his expressed intention to present the material without prejudice or emotion. He has accepted these reports uncritically, although they have not been scientifically substantiated. Furthermore, many of his deductions from these reports are unwarranted. This book would much more aptly be entitled "The Case for $H_3$ Injections."

Suppose I were to take advantage of your assurances that neither you nor Mr. Marx would care how or what I commented, if the net result would be an objective book. Even if I could counteract some of the more obvious leanings of this text, we would still run into a quite different problem. The reader would naturally assume any statements not commented upon to be wholly acceptable. This would rarely be so. To counteract the "prejudice and emotion" which is manifested in choice of words, highlighting of certain statements, etc., would require rewriting the manuscript.

An additional objection is the possibility that the original plan of monitoring might lend an air of "medical authoritativeness," and imply acceptance by myself.

I hope you understand that nothing I have said is meant to reflect on the intentions or the ability of Mr. Marx. It is perhaps impossible to write a book for the layman on a controversial medical subject in an objective manner.

I cannot make any recommendations regarding disposition of this manuscript. I might suggest that you wait until further work by Prof. Aslan or the research into the subject being conducted elsewhere provides a definitive answer. On the other hand, publication of this text, even as it stands, might disabuse those people who think there is some secret ingredient or formula in Aslan's preparation, and I believe might equally disabuse any people who hope to return to the days of their youth. Frankly, I would rather see it published as it is, than with any implication that the text has been "medically monitored."

But both you and the author will have to live with the fact that publication of this book as it stands is certain to generate pressure for premature generalized usage. I feel strongly that any indiscriminate prolonged administration of procaine at this stage of inadequate scientific documentation is to be rejected.

I also want to make it clear that I am not expressing any opinion on the efficacy of the treatment. At this time I have insufficient data to permit acceptance or rejection.

I reiterate—I cannot advise you—evidently publishers, like doctors, sometimes have to make difficult decisions, which no one else can make for them.

———

**Our decision was to publish the text as it stood, prefaced by this vigorously critical statement.**

F. C.

# Chapter 1

*We face the problem*

In the early years when men still ate lizard's tongue and mandrake root as cures, only the strong — the young—were equipped to survive the myriad dangers and diseases rampant upon the earth. (Indeed, in many cases only the young were sufficiently well equipped to survive the cure!) The aged were too slow to flee from fire or flood. They could not slay a wild boar, nor even outlive a witch's curse. They were the easiest of prey in a world in which even the fit could not survive all the machinations of plague and famine. The human race was young. But already the old were outdated.

As time went by, the struggle for mere survival was easier. Man had learned to couple his energy with the resources about him. Now he had other tools with which to fight. He could live out his natural life, and dared even to dream of extending his life span.

(It is interesting to note that almost all the tales which mirror man's dream to live out his natural life

1

with vigor and independence are permeated with an aura of the supernatural or anatural. Did not Faust have to sell his soul to the Devil? What of Daniel Webster's classic debate with Scratch? And the narcissistic Dorian Gray?)

Despite the connotation of evil surrounding such a course, man rejected the image of an unproductive future. Ponce de León searched for the Fountain of Youth in an era in which man still chased myth. (Interestingly enough, some three hundred years later the land on which he sought eternal youth has become a "paradise" for the aged.) The Bible relates King David's request for young girls to be placed in bed with him in order to effect the monarch's rejuvenation. While the scriptures do not elaborate upon the dynamics of such a "cure," centuries later man still believed in the rejuvenating power of inhaling the breath of the young: the famous Dutch physician Hermann Boerhaave recommended such a therapy as late as the 18th century. However, rejuvenistic literature was in its heyday during the middle ages, when superstition was riding high and magicians, alchemists and charlatans had the field to themselves.

Roger Bacon disclaimed any belief in magic. However, his writings indicate that as far as the possibilities for rejuvenation were concerned, he was by no means more enlightened than were his colleagues of the 13th century. Paracelsus was the most famous physician during the 15th century, and while lately some of his ideas have been resurrected, his arcanum for immortality has long been forgotten (perhaps because he died at the age of only 48).

Nothing approaching a scientific attempt at rejuvenation took place until the end of the 19th century. At about the same time that Pasteur experimented with anthrax and William Morton introduced ether anesthesia, a French scientist, Charles Edouard Brown-Séquard, astonished his associates at the Société de Biologie by appearing before them, after several months of seclusion, looking at least twenty years younger than when last they had seen him. Brown-Séquard was a man of 72 who had lost his zest for life—only the scientist in him could not be subdued. He proudly explained to his audience that through the injection of animal testes he had "rejuvenated" himself: his irritability and impotence were gone, his gastrointestinal and urinary problems had diminished. At the same time, his muscular power had increased, which he demonstrated with the aid of an ergograph (a mechanism designed to show graphically the work and fatigue of muscles).

Brown-Séquard, until then a highly respected scientist, soon found himself condemned by many of his peers. The results of his self-experiment were ascribed to his "senile-erotic imagination." Nevertheless, although he had by no means "rejuvenated" himself, he had demonstrated the importance of internal secretions to the vigor and strength of the human body. Unfortunately, his method did not combat old age, but succeeded merely in activating the organism.

Only 17 years after Brown-Séquard's death at the age of 77, and in the shadow of Ehrlich's discovery of Salvarsan, the Viennese physician Eugen Steinach advanced another theory: he advocated tying off the spermatic cords, thus preventing the production of wasted sperm, and increasing the internal secretion which is passed into the bloodstream. He thought aging to be connected with the involution of the interstitial cells of Leydig, cells in the testes which to this day have not been isolated and whose function has never completely been explained. Steinach named the interstitial cells "puberty glands," and proceeded to effect a "surgical reactivation" of the male by so-called vasoligation. The operation, performed under local anesthesia, was a difficult one which provided only temporary benefits. Today, Steinach's operational technique is used only occasionally, in cases of urinary complications.

While the Scottish bacteriologist Alexander Fleming was growing mold cultures in a search that culminated in penicillin, Dr. Serge Voronoff, soon after World War I, stirred up a great deal of curiosity by grafting monkey sex glands on humans. Again, the effects were not lasting. Voronoff himself claimed only that he could prolong the vigor and joy of life for five to six years by this method, after which one more grafting operation (not entirely without danger) was feasible. However, should the individual survive twelve years, he was doomed to hopeless senescence, which perhaps made things worse than they were before.

Voronoff was, in turn, followed by the Russian physician and biochemist Aleksandr A. Bogomolets, who developed ACS (antireticular cytotoxic serum), based on his premise that stimulation of the physiological system of the connective tissue was of great importance in preventing morbid aging. His serum was supposed to retard the gradual exhaustion of the body, thereby delaying the onset of senility. This theory enjoyed a brief vogue, but when Bogomolets died (at the age of only 65), most of the ardent supporters of his theory turned their attention to the newly discovered sulfa drugs.

Was it a quest for personal immortality that stimulated Brown-Séquard or Steinach? Voronoff or Bogomolets? Who can tell? We do know that the time was ripe for the discovery they were seeking. Medical science had already made great advances. Men over forty were no longer considered old, and those over sixty no longer so rare as to be venerated for their age alone. Painful, helpless old age was soon to become a major medical and social problem, but these first scientific attempts to preserve true life in the aged unfortunately held the attention only of sensationalists and fanatics. The attention of medical scientists was still directed to the most pressing medical problem—the control of infectious diseases. The goal of *less disease* had to be reached before the problem of *healthy longevity* could command widespread research attention. Furthermore, the substrate of biological knowledge essential for true progress in this field was not available at that time.

Unsophisticated as were these pioneer attempts by scientists to preserve vigor throughout old age, they nevertheless presaged one of the most ironic dilemmas of modern times. We have learned how to keep ourselves alive to a ripe old age, but we have not learned how to make this old age worth the living.

Now that the scourges of bubonic plague, smallpox, malaria, typhoid, yellow fever, and polio have been checked, and we have learned to use vitamins and antibiotics, blood banks, and new surgical techniques, we are confronted with the success of our toil. The number of people over 65 doubled in the United States between 1930 and 1950. In 1958 there were *more than 15 million people over the age of 65* in this country (a 23 per cent increase over 1950). And today the 65-year-old has a life expectancy of 14.1 years. Having successfully prolonged life for so many, we can not afford to stand helplessly by while they and ourselves become prey to the multitudinous complications of old age.

Once the aged were victims of man's weakness and ignorance. Today they are the victims of his knowledge. They are alive—but they cannot flee from their loss of independence, they cannot fight the diseases that plague them, they cannot outlive the curse of senility. The old do not die as easily today; they linger, they whimper. The strong must still carry the weak— and carry them for a longer time.

# Chapter 2

*What is old age?*

Different manifestations of old age have been cata-
logued for many years, but scientific attention was not
directed to them until a half-century ago. A New York
physician, Ignatz Leo Nascher, coined the word
"geriatrics" (from the Greek *"geron,"* meaning old
man) and applied it to that special branch of medicine
which deals with the typical diseases of the advanced
years. The word "gerontology," meaning "study of old
man," was derived, naturally, from this root. Thus,
while the former (geriatrics) denotes the practice of
medical means to alleviate the results of aging, the
latter term (gerontology) categorizes the theory of
aging.

The most perplexing questions facing medical science today concern the causes of old age and the precise measurement of the process of aging in the human
body. We are all familiar with the manifestations of
age, but we are not sure whether its signs and symp-

toms are inevitable, as believed until quite recently, or whether they can perhaps be postponed, or even completely prevented. The growing number of biologists and physicians who advocate the latter believe that the signs of old age are merely indications of a sickness which should yield if properly treated. They point out that chronological and biological age do not necessarily coincide—that some people, in appearance, attitude and behavior, seem younger than they really are, while others appear older. Why this should be so has never been answered satisfactorily.

The aging process actually begins at birth, but at the onset these changes lead to growth and maturity. Once the latter stage has been reached, a slow decline sets in—affecting different parts of the body at different times, sparing some organs until a fairly advanced age. Thus, it has been found that hearing is best at the age of 10, vision at 20, muscular strength and coordination at 25, and reproductive functions at 30. These few examples suffice to show that aging is a slow process, and that it does not affect the entire organism at any single specific time. We speak of "normal aging" and "accelerated aging;" but although our yardstick for the former is still rather vague, there is general agreement that the latter is due to morbid changes in the body.

While the true causes for the decline of vigor in man are not known, we do know that the symptoms which accompany his slow but steady decline usually gain momentum during the fifth decade of his life. The first evident changes are connected with his appearance: hair becomes sparse and grey, shoulders develop a slight stoop, the skin shows wrinkles and the gait loses its earlier buoyancy. Man begins to "look old," although at this stage of the aging process he may not feel old. In fact, because he maintains his activities at the high level of youth, without much change in his way of life, he may well be laying the groundwork for future illness. (An investigation into the dynamics of "middle-age medicine" is another program still to be developed.)

Other symptoms enter the picture: the metabolism is less active, recuperative powers slow down, the energy used up by the body is replaced at a much slower rate, thereby rendering the tissues of the individual organs more flaccid. The skin becomes thinner and, due to a reduction in the secretion of some of its glands, drier. As the years progress, muscles fail to maintain the body in its former erectness, motions lose some of their power and sureness. The thorax is no longer expanded as before, partly due to increased muscle weakness, partly to an ossification of the costal cartilages. Thus, the lungs do not receive as much oxygen, the walls of their air cells become thinner, breathing is less thorough, and the respiratory metabolism slows down.

Bones become more brittle, they decrease in volume and weight, almost all cartilages lose their elasticity and, in certain spots, calcify. A great many of the capillaries in which blood is transported to distant parts of the body become clogged, preventing an even blood supply (the reason for circulatory troubles in the extremities, as well as opacity of the cornea, which leads to cataracts). The digestion, too, becomes impaired; teeth are no longer as efficient as they formerly were; liver, gall bladder, spleen and pancreas are weakened; the absorption of essential vitamins and minerals from the intestines is reduced. The slowing down of the metabolism as a whole also affects the production of heat in the body; older people become extremely sensitive to cold weather. If the aged are not well fed there is a great danger of rapid decline, since the body is no longer able to replace the missing dietary elements from its own storehouse.

Increased atrophy of many organs finally affects the brain, too: as the amount of water in its cavities increases, it begins to lose weight. Mental powers decline (much later, however, than physical abilities), and the loss of memory and orientation, sometimes imbecility, and often a childlike behavior are characteristic. The limbs stiffen, steps become shorter and the climbing of stairs is a problem. The body loses its alertness: once in a quiet position, it has to be brought into motion rather slowly. Gestures are no longer as frequent as before, and the older man finds it difficult to make different motions simultaneously: if, for example, he has to button his coat on the street, he will stop walking. His coordination is further complicated by a shaking of the limbs.

In addition, senility is now one of the most frequent and pathetic manifestations of old age. It is also one of the most expensive, but its cost can not be reckoned in dollars and cents alone. Senility is not so much the despair of its victims (who usually are oblivious to their fate) as of their families, who must bear the physical and emotional strain of caring for the senile parent in the home, or the financial strain and feelings of guilt attached to care outside the home. Just as we have no clear-cut definition for "old age," we have none for senility. Again, we are able to recognize its symptoms, but to go no further. Its mildest signs are a decrease in mental acuity and an increase in forgetfulness; the severest, psychosis and regression to infantile behavior. Between these two extremes we find apathy, confusion, hostility, restlessness, sloppiness, asocial attitudes, sexual aberration, and persecution feelings.

Why some people become senile and others do not is still unknown. Social and psychological stresses play a part, but of primary importance are the changes in the body. Hormonal disturbances, a malfunctioning enzyme system, and lowered activity of the brain cells (stemming from a reduction of the cerebral blood flow) seem to be the main causes of man's mental deterioration.

Dr. William Malamud of Boston University recently has stated that the incidence of mental disorders in old age has been rapidly increasing for the last three decades and, particularly since the end of World War II, "has skyrocketed to a degree totally out of proportion to any of the other types of personality disturbances."

# Chapter 3

*Why do we age?*

The signs of old age are no more profuse than the theories attempting to explain its cause or causes. This is hardly surprising, for no common denominator has been found as yet. There is not even agreement as to how many years constitute man's natural life span. Each plant or animal has a maximum life expectancy, which cannot be exceeded even if the organism remains healthy. The fact that almost no one dies of old age per se would indicate that most of us never reach our full potential. But what is our potential life span? Biologists measure it against the time maturity is achieved or bones stop growing in length, and their resulting figures are that man should live for 120 to 150 years! But no generally accepted criteria exist: man's natural life span is still a matter of speculation.

Aging, according to one school, is due to an exhaustion of life energy; another holds that it is due to a slowing down of the metabolism (although it may be questioned whether this is not the consequence rather than the cause). A third group maintains that the flooding of the organism with toxins is responsible for aging; Metchnikoff, one of the first scientists to occupy himself with this problem, was convinced that auto-intoxication caused old age, and that death followed the accumulation of fatal toxins in the large intestine. A more mechanical hypothesis, which was developed many years before the first atomic explosion, deals with the possible effects of cosmic radiation on the life span. The involution of individual organs (i.e., sex glands, ovaries, thyroid or other endocrine glands) was blamed for aging by a school which believed that renewed vigor could be conferred on old people through a genuine reactivation of these glands. Also, the degeneration of the nerve cells, in particular a creeping paralysis of the central nervous system is considered by some as the primary cause of aging.

One of the more widely accepted theories of aging is that of Dr. Hans Selye of Montreal. The originator of the concept of stress diseases, he defines stress as the "rate of wear and tear in the body." This wear and tear is a continuous process and has a cumulative effect. According to Selye, each human being (or animal,

for that matter) has only a certain amount of "adaptation energy" with which to replenish his vital reserves.
Theoretically, this reserve shrinks a little after each
stress, and the deficit in adaptation energy, occurring
from day to day, "adds up to what we call aging."

Among the first signs that the body is failing to
adapt itself to stressful situations are the many allergic
diseases, such as hay fever, certain rashes, and asthma.
And the diseases of old age are for the most part
caused not by invasion of the body by germs or viruses,
but by failure of one or another part of the body to
adapt to the stress of life.

To quote Dr. Selye: "Among all my autopsies (and
I have performed quite a few) I have never seen a
man who died of old age. In fact, I do not think anyone
has ever died of old age yet. To permit this would be
the ideal accomplishment of medical research (if we
disregard the unlikely event of someone discovering
how to regenerate adaptation energy). To die of old
age would mean that all the organs of the body would
be worn out proportionately, merely by having been
used too long. This is never the case. We invariably die
because one vital part has worn out too early in proportion to the rest of the body. Life, the biologic chain
that holds our parts together, is only as strong as its
weakest vital link. When this breaks, no matter which

vital link it may be—our parts can no longer be held together as a single living being."*

Since living cells in a watery solution can be kept healthy for an infinite period of time by cleaning them and returning them to a fresh solution, Selye advances the idea that the weaknesses of old age may be due partly to an accumulation of waste products which interfere with the nourishment of the cells.

If this is the mechanism of aging, Dr. Selye points out, there should be at least two ways of avoiding it. The rate of waste production might be slowed down, or the system might be helped to destroy its waste and get rid of it. Research on this and other approaches to the causes of aging is now being conducted by Dr. Selye and his staff at the Institute of Experimental Medicine and Surgery in Montreal, which he founded in 1949 and has directed ever since.

Dr. Selye believes that medicine has now assembled a fund of knowledge that will serve as a point of departure for studying the causes of old age. Again in August of 1959, he reiterated his conviction that aging may be regarded as a disease and . . . "like any other disease, it is probably preventable or curable . . . The truth is that death by disease is largely avoidable."

He is not alone in this belief.

*The Stress of Life. McGraw-Hill, New York.

# Chapter 4

*Parhon—a pioneer*

One year after Nascher had coined the word "geriatrics," and while Hans Selye was still a schoolboy, a young Rumanian physician named Constantin I. Parhon took the position that old age is a treatable disease. The first Rumanian medical figure to achieve international renown, Parhon, now a man of 85, began his career as a neurologist but soon became attracted to the functions of the endocrine glands. On the basis of experimental and clinical research he began administering an extract of the pineal gland (a small gland attached to the posterior of the brain) to old people. While his method was much simpler than that of Steinach, his successes were by no means more definitive. However, continuing his endocrinol-

18

ogical studies, he found that animals would develop signs of old age if certain glands or tissues—such as the thyroid, the spleen or the mammary glands—were removed surgically. His treatise on the importance of the endocrine glands, published 50 years ago, is considered the first large scale endocrinological work in medical literature.

For 40 years Parhon continued his investigations into the causes of aging. His aim was to determine the true biological age of human beings. During 22 of those years, Parhon was a Professor of neurology, and it was not until 1934 that a chair of endocrinology was created for him at the University of Bucharest. Six years later the Rumanian Fascist government fired him, but he returned to his former position in 1945. It was soon thereafter that he founded the Institute of Endocrinology.

Professor Parhon's views can best be summarized in his own words from one of his more than 1,200 publications:

"From a theoretical point of view, I am of the opinion that aging begins simultaneously with growth and development, and that the mechanism of aging can be understood only in terms of research concerned with the changes which the entire organism undergoes throughout its lifetime. The phenomena that determine and accompany the aging process are so numerous and involved and their mode of origin so deserving of study that they cannot but fascinate all biologists . . .

"I am of the view that the process of aging occurs only to the extent that the conditions giving rise to it have occurred. If one were able to interfere with the mechanisms of aging, the direction of this process would be subject to change. In this way it is conceivable that the aging organism, whether the aging is due to premature, pathological, or so-called normal factors, could be returned to an earlier biological state. My experiences . . . have shown that the rhythm of life can be either speeded up or slowed down at all stages. . . . Biological and chronological age are not necessarily identical. Differences in aging rates can also be observed in clinical situations, e.g., endocrine conditions, and I believe that the rate of aging, 'the film of life' as it were, can be controlled in either of two directions, i.e., toward faster aging or, to some extent, toward rejuvenation.

"If one were to view the aging process as irreversible, steps to control it would involve merely sanitation and the usual treatment. But if the aging process is regarded from a functional viewpoint, as a deviation from the normal functional optimum of the individual, i.e., as an abnormal phenomenon, then treatment no longer seems impossible. In our view the aging process is a pathological condition, or, to state it better, a more or less extensive dystrophy . . . which develops slowly as the organism grows and differentiates. It is our obligation to treat these disturbances of function, and to prevent them whenever possible."

Parhon's interest in the problems of aging (in 1926 he coined the word "ilikibiology" from the Greek "*iliki*" meaning old age, but this never gained wide acceptance) led, in 1951, to the founding of the Institute of Geriatrics in Bucharest. He then had at his disposal facilities which made it possible for him to pursue vigorously his important work. The nucleus of the Institute was an existing old age home with almost two hundred inmates. These were the people through whom Parhon wanted to prove definitely that old age could be treated. All of the inmates suffered from serious degenerative diseases; he proceeded to give them either tissue extracts (of the spleen or placenta), gland extracts (adrenal and pineal glands, thyroid), vitamins (Vitamin E, liquid beer yeast), or baths of bicarbonate of soda (a treatment developed by the Russian physiologist Olga Lepechinskaya, which had received broad publicity in the Soviet Union in the post-war years). All of these treatments had been tried previously at other institutions, and the results in Bucharest were about the same: a few of the old people seemed to benefit, but no really important changes took place.

In 1949, a woman doctor who for more than 20 years had been a specialist in cardiovascular diseases had joined the staff of Dr. Parhon's Institute of Endrocrinology. Prof. Anna Aslan had also long been interested in the pharmacodynamical properties of novocain* and its action in the human body, and she continued her experimentation under Dr. Parhon's direction—and under the encouragement of his conviction that old age and its manifestations are treatable and preventable.

It would seem that Dr. Parhon was impressed with the capabilities of his new staff member, and with the direction and progress of her research, for in 1952 Parhon, now honorary head of the Institute of Geriatrics, turned over its directorship to Prof. Anna Aslan.

---

*The word Novocain is a trade name, owned by Winthrop Laboratories, for procaine hydrochloride. Like the word Adrenalin, which is also a trade name, novocain has become an accepted word in the English language. It is used in this book with the permission of its owners.

# Chapter 5

*Rejuvenation—an emotionally charged word*

The story of Prof. Anna Aslan and the therapy for the diseases, discomforts, and agonies of old age developed by her and her co-workers at the Parhon Institute of Geriatrics is an exciting one, it has also been frequently misrepresented, and even when reported objectively, this story is easily misinterpreted.

When we are children, we can hardly wait to be grown up. At some point, we reach that very satisfying period of life when the privileges and pleasures of maturity are combined with an eager zest for life and with the vigor of youth. This delightful stage—usually the late teens to the late twenties or early thirties—goes by most of us unnoticed and unappreciated (how often do the middle aged and older quote the phrase "youth is wasted on the young"). Then some little sign —a hair line thinner at the temple, some wrinkles around the eyes—reminds each of us that, like all living things, we too are subject to the aging process. And one day, for the first time, the thought flashes through our mind "If I were young again . . ."

This thought is father to the wish. But if we no longer believe in the curses of witches, neither can we hope for the magic wand of a fairy to touch us and restore youth and beauty. There is no way to turn the clock backward. So most of us go on about our living, the wish to be again young (or at least middle-aged) buried in our subconscious.

The word *rejuvenation*, expressing this buried wish, is perhaps charged with more wishful thinking than any other word in our language.

For this reason it is unfortunate that Prof. Aslan has used the words 'true rejuvenation' in describing occasional effects of the therapy she has developed. For her words have been picked up by headline writers and newscasters, and shortened into such deplorable catch-phrases as "Youth shots," "Youth serum," "Long-life drug," and even "Live-forever juice."

> Webster's unabridged dictionary defines *rejuvenate* as follows:
>
> —To render young or youthful again; to impart renewed vitality to.
>
> (Med.) To restore to a more youthful condition; specif., to restore sexual vigor, as by an operation.

Professor Aslan has used the term rejuvenate in the restricted sense of "to impart renewed vitality to . . . to restore to a more youthful condition." It is *only in this sense* that I shall use the word rejuvenate subsequently in this text. In this limited sense, rejuvenation is a phenomenon familiar to us all. Our vitality is renewed when we convalesce from a serious illness, or vacation after a long period of strain or over-work. Surely we are restored to a more youthful condition as we recover from the shock and grief of the loss of a dear one, or have some great worry lifted from our shoulders. In fact, nature imparts renewed vitality to us and restores us to a more youthful condition each time we enjoy a good night's sleep.

> **Nowhere in any of her published reports or public statements has Prof. Aslan claimed to be making old men and women young again, and any restoration of sexual vigor which has occurred in any of Prof. Aslan's patients has been incidental to the renewing of the vitality of that patient.**

With these facts understood, let us now see on what basis this brown-eyed, dark-haired woman doctor in Bucharest claims that a treatment she has developed "imparts renewed vitality to old men and women" and "restores them to a more youthful condition."

# Chapter 6

## *A new use for an old drug*

Anna Aslan was born in Bucharest in 1898. Her first ambition was to be an aviatrix, but her middle-class parents were able to dissuade her. They were not so successful with Anna's second, but perhaps more mature, choice of a life work. When they refused to give her permission to enroll as a medical student, Anna Aslan went on a hunger strike. Four days later her family gave up. In 1924, Anna Aslan received her doctor's degree at the University of Bucharest.

For sixteen years, she was an assistant at the Second Clinic in Bucharest under Prof. D. Danielopolou, whom she reveres as her teacher. She held positions in other hospitals as well, and, except for three years during the post-war period when she was director of the Clinic in Timisoara, has always remained in the Rumanian capital.

Immediately after World War II, Prof. Aslan learned of a therapy which had been pioneered by three French physicians, Dos Ghali, Bourdin and Guiot. The doctors had injected novocain into the cubital vein twice within two hours in an attempt to effect relief in patients suffering from asthmatic attacks, and their method was successful where others had failed. Prof. Aslan found that she, too, could help asthma sufferers with repeated injections.

Then, in 1948, the noted German physician Prof. Gustav Spiess died. He had been first to discover that novocain not only has value as a local anesthetic, but it also has curative power. Prof. Aslan read his obituary in a Rumanian medical journal which noted his former achievements. After checking through the literature, she immediately extended her novocain treatments to include patients with arteritis and limb embolisms, using the method devised by René Lériche, which even went further than Spiess' original idea. Lériche had advocated the infiltration of 10 to 25 cc of novocain, and was able to restore the affected joint or limb to full activity, often after as few as two treatments. Prof. Aslan, encouraged by her first results with novocain, began to use it also in cases of arthritis and arthrosis with a tendency toward the fixation of a joint (ankylosis).

The efficacy of the treatment was confirmed in the very first test:

> "G. J., a medical student, came to us with arthritis of the right knee, having had severe pains in his knee for three weeks. After intra-arterial injection of 0.10 g of 1 per cent novocain, he was immediately able to flex his joint up to 90 degrees."

Before proceeding on a larger scale, she thought it best to experiment with animals. Dr. Selye had already reported a method for inducing an experimental arthritis in mice. When a drop of some irritant solution, such as formalin or croton oil, is injected under the skin of the sole into one of the hind paws of a rat, a local experimental arthritis develops. First there is acute swelling at the site of injection, and this swelling gradually transforms itself into a chronic arthritis of the many small joints in the paw, and especially of the ankle joint. The rat becomes permanently crippled, because the joints stiffen with hard connective tissue, so that they can no longer be moved.

In the course of her experiments on mice in which arthritis had been induced by a slight modification of this method, Prof. Aslan and her coworkers at the Parhon Institute of Endocrinology (which Prof. Aslan had joined in 1949) not only found that the novocain had a therapeutic effect—they also observed that the treated animals gained weight, and developed a lustrous fur. Complete cures were achieved in 85 per cent of the affected animals, and resistance to the experimentally induced arthritis was greater in the prophylactically treated animals.

After this series of successful animal experiments, she began treating a group of selected patients between 1949 and 1951. Not all of them were helped, but improvement in many cases was gratifying to the doctors at the Parhon Institute, for they had not achieved really effective results with any other method.

While proceeding with these treatments, Prof. Aslan made a most important observation: the patients, in her own words, "showed a change in the psychological and physical conditions, an improvement in memory, a decrease in rigidity due to Parkinson's disease, and an increase in muscular power." Prof. Aslan then checked through the literature again. She could not find a single reference to any such effect of novocain. However, she had witnessed those changes with a trained, professional eye — the patients appeared younger than before, much more alert, and seemed to be enjoying life again.

Of course, it was possible that these changes might be due merely to the cessation of pain, and the fact that hope had replaced the depression and resignation which had consumed the patients. There was still no proof that any physico-chemical changes were occurring in the bodies of the novocain-treated patients which would not have occurred as the result of balanced diet, normal regime, and 'tender loving care'.

Professor Aslan then selected 25 of the patients, and for three years treated them with novocain, while all the others continued to receive only gland extracts, vitamins, etc. The apparent greater vitality and improvements in specific diseases of the novocain-treated group, as compared to the others, were convincing to the Professor and her co-workers, although they still did not detect any significant physico-chemical changes.

In 1955 she published her findings in the Journal of the Rumanian Academy of Science (*Bulletin Stiintific Academia Republicii Populare Române*). This paper, entitled "La novocaine, facteur eutrophique et rejeunissant dans le traitement prophilactique et curatif de la viellesse" (Novocain — a eutrophic and rejuvenating factor in the prophylactic and therapeutic treatment of aging), contained a report of her work on the "25 cases."

As might have been expected, her Rumanian colleagues who knew of her work but had not seen its results considered it nothing short of preposterous that she make such fantastic claims. Few non-Rumanian doctors follow the proceedings of the Rumanian Academy, and thus the knowledge of this therapy was confined to its country of origin. But, fortunately, not for long.

Shortly after the Academy report was published, a German journalist traveling through Rumania paid a visit to the Institute, and included a paragraph or two on its work with old people in his articles. Those few sentences attracted the attention of Farbwerke Hoechst, a West German drug firm, which had been the sole manufacturer of novocain in Germany since its discovery. This unsuspected new use for novocain intrigued the company, which several years before had issued a booklet detailing the therapeutic versatility of this substance. Doctor Horst Weeke, medical director for Farbwerke Hoechst, was sent to Bucharest to investigate the matter. Upon returning to Frankfurt, Dr. Weeke wrote a very positive report on Prof. Aslan's accomplishments. This report was not published, but instead was circulated among some quite important German medical men. As a result, Prof. Aslan received an invitation to attend the Karlsruhe Therapy Congress (one of the best known German medical conclaves) and to read a paper on her discovery.

September 3, 1956 was the date on which Prof. Aslan faced her first Western audience. The two dozen people listening to the obscure Rumanian doctor were skeptical and filled with incredulity at her tales of the apparent rejuvenation of old people. The applause she received was thin; indeed, the doubts which hushed the congress could be seen on every face, and some even spoke openly of a "great hoax." The meeting was concluded with the sentiments that Prof. Aslan's remarks did not belong in the program of a reputable medical congress.

# Chapter 7

*Professor Aslan's claims for "H₃" therapy*

What had Prof. Aslan actually said to engender such skepticism?

She had described her experiments and their results, in the 25 original cases, and the 45 old people who were being treated at that time as residents of the institute. She cited some specific case histories from the original group of patients, and summarized her observations of the results of "$H_3$" therapy in both the in-patients and 2500 oldsters who were being treated on an out-patient basis, as follows:

"Our initially cautious and limited investigations, originally restricted to twenty-five cases, have been expanded appreciably in the last two years, thanks to the innocuousness of the substance involved (novocain) and to the successes gained.

"*Clinically* we have achieved a reversal of phenomena which until now have been considered irreversible, e.g., in cerebral arteriosclerosis. Sustained improvement was achieved in cases of senile Parkinsonism, improvement in hearing, and in certain reflexes which could not be elicited

33

prior to treatment. Also, renewed production of estrogens and of characteristics resulting from stimulation of testes and adrenal glands, retrogression of signs of senility of the skin, such as ichthyosis and senile keratosis, repigmentation of existing hair, stimulation of new hair growth, and fewer arteriosclerotic symptoms.

"*Physiologically* we found improvement of the central activity of the nervous system, improved cardiovascular reaction to stress on the part of old people, decreased oxygen consumption, better muscle power.

"*Biochemically* we found alterations in protein structure and in the ratio of albumins to globulins, increase in cholesterol (as compared to lower than normal values prior to initiation of treatment).

"*Hematologically* we found fewer leucocytes, higher production of granulocytes, increase in the number of monocytes and the globulin content.

"*Eutrophic action* was clearly visible in cases of atrophic ulcer, stomach ulcer, dermatosclerosis, psoriasis, rashes, alopecia areata and leucoderma.

"It can be stated that novocain reduces the biological age of those treated with it below the chronological age. Novocain affects directly the cerebral cortex and its dynamics, and acts on the whole nervous system, the diencephalon centers, the spinal cord, peripheral nerves, metabolic processes, and brings about trophic changes in the entire organism. It also affects the endocrine glands. Through its vitamin-like action (due to its content of para-aminobenzoic acid) it also acts favorably on the biocatalysts, and it seems to stimulate the intestinal flora to the production of biogenic agents.

"Its trophotropic action can also be observed in arteriosclerotic processes. Its effect in mobilizing the cholesterol of the blood vessel walls may be due to its hydrotropic action, characteristic of the chlorides of para-aminobenzoic acid.

"Novocain minimizes the feeling of sickness and leads to a heightened desire and capacity for physical and mental activity. By virtue of its trophic action and its role as stimulator in most vital processes, it may be considered a useful prophylactic and therapeutic substance in the fight against old age."

Why did the attending doctors find these words impossible to believe? First, undoubtedly, because the word "rejuvenation" brought back memories of the Brown-Séquard fiasco, and they had not listened attentively enough to realize that Prof. Aslan was by no means claiming that her therapy would make old people young again. Yet even her modest claim ". . . a useful prophylactic and therapeutic substance in the fight against old age . . ." seemed too good to be true.

Her suggestion that novocain be termed "H₃" may have led some inattentive listeners to believe she claimed to have discovered a new substance. And, finally, there was still no scientifically unassailable proof that even if all Prof. Aslan reported was true, novocain deserved the credit. We will discuss in a later chapter the type of rigidly controlled experiments which will be necessary to prove finally or disprove the efficacy of novocain injections in old age.

# Chapter 8

*"H₃" = novocain = procaine*

Perhaps the closest comparison to the claims Prof. Aslan made for novocain lay in fiction: the drug "Soma" induced a sense of bliss with small doses, brought visions with larger doses, and, finally, refreshing sleep to the inhabitants of Huxley's "Brave New World." The history of novocain is almost as interesting.

Local anesthesia with cocaine was introduced by Carl Koller in 1884. Koller, a Viennese opthalmologist, had his attention directed to this drug by Sigmund Freud, then a young doctor. Cocaine had been extracted from the South American coca leaves some twenty-five years earlier, but its importance remained uninvestigated. When Koller used it, first on the eyes of frogs and then on human beings, he marked the beginning of local anesthesia. Freud, who originally had the idea, was given no credit for his foresight.

There was no question concerning the efficacy of cocaine. Once put into practice it altered many surgical procedures. But it was also clear that this was a highly toxic substance, and, furthermore, it created a feeling of euphoria which some patients valued highly. Many of them became cocaine addicts, thus inverting any benefits they might have received from its use. In 1905, Alfred Einhorn, a German, produced procaine hydrochloride (the generic name for novocain) by combining para-aminobenzoic acid and diethylamino-ethanol. While there is no chemical relationship between cocaine and novocain, the latter has the same effect as the former except that it has a considerably lower toxicity and is nonhabit-forming. Many other substitutes for cocaine have been synthesized since then, but novocain has remained the most widely used.

The novocain with which most people who have had teeth drilled or extracted are familiar is the hydrochloride of procaine. This hydrochloride has the chemical formula $C_{13}H_{20}O_2N_2HCl$.

The structural formula of procaine is

O=C—O—CH₂—CH₂—N(C₂H₅)₂·HCl

... a comparatively simple compound when compared to the complex pharmaceutical substances which have been synthesized in recent years.

Novocain consists of small, colorless, tasteless crystals, easily soluble in water. It is generally well tolerated, and up to a concentration of 10 per cent, produces no noticeable irritation of the tissues. Its toxicity is much lower than that of cocaine. While the latter is excreted slowly, novocain is hydrolyzed within 20 to 30 minutes and detoxicated by the blood and liver. The toxicity depends, therefore, less on the dose than on the route and speed of injection or infiltration. There is never any addiction to novocain, nor does the body develop a tolerance which would make larger doses necessary in order to bring about the desired results, as is the case with cocaine or morphine.

Novocain is hydrolyzed with the help of an enzyme, called procainesterase, which is present in some of our tissues. The products of breakdown — the same from which the novocain was made — are para-aminobenzoic acid (actually a vitamin) and diethylaminoethanol, both considered virtually nontoxic. When, however, efforts were made to have some of the pharmacological properties of novocain explained through the use of its products of hydrolysis, researchers were time and again frustrated by the fact that the effects which they tried to produce eluded them. Continued experimentation confirmed the theory that, in dealing with novocain, the whole is more than the sum of its parts.

Prof. Aslan also attempted to produce the same effect with the vitamin component of novocain—its para-aminobenzoic acid. This attempt was not successful, and because she did not (and still does not) know exactly what part or parts of the drug novocain caused the revitalization and reinvigoration she had observed, Prof. Aslan and her co-workers in Bucharest gave this unknown factor the name "$H_3$," to distinguish it from its component vitamin, which is a member of the B-complex of vitamins and is known as $H_1$, and from folic acid, which is also a B-complex vitamin and is known as $H_2$.

When news of Professor Aslan's therapy began to make headlines in England and the United States, newspaper reporters found it much easier to fit "$H_3$" into a headline than either the familiar novocain, or the precise chemical name—procaine. An impression was thus unfortunately created that the Rumanian doctor had discovered some new drug, or that some mystery ingredient had been activized. Nothing could be farther from the truth. To put it most simply,

$H_3$ = novocain (a trade name) = procaine.

Subsequently we shall refer to this drug as procaine,

in accordance with the preference of American and British doctors and biochemists.

Not only has no new drug been discovered in Bucharest—Prof. Aslan, modestly but with more than a modicum of truth, terms her applications of procaine in the diseases of old age *"a rediscovery."*

# Chapter 9

*Why a "rediscovery"?*

At first, procaine was employed exclusively as a local anesthetic, but physicians soon began to discover that this drug also had certain therapeutic effects. Even before the advent of procaine, Karl Ludwig Schleich showed that certain rheumatic diseases and muscle pains could be controlled with the infiltration of cocaine. However, his reports were largely ignored. In 1906, Gustav Spiess pointed out that inflammations could be rapidly reduced with procaine injections.

> "An inflammation will not break out if we succeed in eliminating the reflexes in (certain) nerves starting from the focus of infection, through anesthesia. An inflammation already existing is cured rapidly by anesthetizing the focus of inflammation."

Other researchers confirmed Spiess' observations, but little use was made of procaine to influence specific diseases via the nerves. It was found that injections and infiltrations of the ischiatic nerve would stop very painful sciatica attacks, that whooping-cough fits could be controlled by anesthesia of the upper larynx nerve, that lumbago of gynecological origin, muscle spasms, tetanus and muscle atrophies could be influenced by use of procaine. But these were mostly observations, some of them very interesting, none containing any conclusive information.

One practitioner who has used procaine for more than thirty years, most of the time in a combination with caffeine, is Dr. Walter Huneke of Stuttgart, Germany, who with his brother Ferdinand, of Düsseldorf, was the originator of the so-called "neural therapy," which seeks the cause for most illnesses in foci, and through injections into those foci (inflamed tonsils, bad teeth, operation scars, etc.) is able to clear up a surprisingly great number of morbid conditions. Several years after Prof. Aslan's first report on her findings about the general changes procaine brings about in older citizens, he said:

> "I was struck by these rejuvenating effects (of procaine). I consider as rejuvenating characteristics restoration of a youthful and fresh appearance, better posture, improvement in skin turgor, in sight and hearing, no more falling of the hair, increase in cerebral functions, such as thinking capacity, memory, sleep, mood, efficiency and elasticity, the increase of sexual and other hormonal functions (return of libido, normalization of menstruation), also an improvement in heart and circulatory disturbances, blood pressure, arthrotic complaints and numerous other manifestations of

old age, some of them measurable through the electrocardiograph, blood pressure or metabolism measurements."

Back in 1952, Dr. Huneke in a published report had commented, almost as an aside, that repeated treatments with injections of procaine and caffeine had in many cases "a clearly rejuvenating and therefore life-prolonging effect." But he did not pursue this course of investigation. This was the job then, left to Prof. Aslan and her co-workers. (In 1959, Dr. Huneke, in collaboration with Dr. Berthold Kern published a book, "*Verjüngung durch Novocain*,"* in which he confirms his earlier assumptions, and takes issue with Prof. Aslan because he considers her injection technique as being too unspecific.)

Procaine has been applied in the following ways:

1. Intravenous injections of 5 to 10 cc of a 1 per cent solution;
2. Drop infusions of 20 minute duration;
3. Local infiltration of ½ to 2 per cent solution in doses up to 100 cc;
4. Blocking of the ganglion stellatum by injections;
5. Blocking of the sympathetic nervous system;
6. Intramuscular injections;
7. Intra-arterial injections;
8. Various anesthetic procedures.

---

* *Rejuvenation Through Novocain.*

To list the more than 150 diseases and afflictions against which doctors have reported the use of procaine by one of the methods enumerated, and with results ranging from complete failure to significant success, would make a tiresome catalogue. Positive results from the use of procaine have been reported in conditions as far apart medically as hypertension and frostbite, colic and fractures, certain eye diseases and heart arrhythmias, some skin disorders, and migraine headache. Its application has been largely on the basis of providing prompt relief from pain, but its therapeutic efficacy has frequently been commented on in the literature.

One of the foremost pharmacologists of our time, Professor Fritz Eichholtz of the University of Heidelberg, reached the conclusion that "of old, honest novocain there has become in modern pharmacology a medical substance which seems destined to be the kind of cure-all so many practitioners dream about." When he made this statement (qualifying it by saying that most of the procaine effects are weak and need an additional specific agent) the work of Prof. Aslan had just begun and was not known outside of Rumania.

Pharmacologists, clinicians, practitioners and, lately, gerontologists, have cleared up a great many of the mysteries surrounding procaine. However, as its range of applications indicates, there is much work to be done before we even approach the limits of procaine's potential; no definitive conclusions have been reached at this time. Indeed, the more that is discovered about this substance, the more confused the issue becomes. The bibliography on procaine comprises today close to five hundred publications: an immense wealth of information has been spread before the medical profession. Still, there are only a few things concerning this drug with which the entire profession is in agreement.

Despite the fact that procaine is nonhabit-forming and has been used in the past for treating dozens of diseases, many physicians have been discouraged from using procaine due to reports of its side effects: nausea, vertigo, difficulty in breathing, vomiting, or temporary visual disorders have occurred in some cases, and strong allergic reactions (where an allergy test was omitted) could produce undesirable complications leading perhaps to anaphylactic shock and death. Most of the negative effects have occurred during spinal anesthesia; investigations have shown that most frequently in those cases either the dose was too high, or else instead of merely procaine, mixtures with cocaine, pantocain or adrenalin had been used. These regrettable errors have prejudiced many doctors against its use (outside of the field of local anesthesia), and the use of procaine had been on a steady decline since 1952.

Before Prof. Aslan began her experiments, procaine had never been systematically applied as a general therapeutic measure over a long period of time. She proceeded to do just this, and refused to let herself be discouraged by occasional failures. A slight modification or "buffering" of the plain procaine, increasing its acidity, eliminated the occasional tissue necroses which Prof. Aslan had observed in its steady use, made the substance more stable, and seemingly speeded up the reactivity of the body.

# Chapter 10

*Some confirming reports*

In the two years that have passed since the 1957 Karlsruhe Therapy Congress, news of the procaine therapy has spread around the globe. Confirmation of the effectiveness of procaine as a treatment for many of the diseases of old age has begun to accumulate. More and more of the original skeptics have begun their own clinical trials of this therapy, and reports of their work appear in increasing number.

### Switzerland

In November, 1957, a group of Swiss gerontologists — Gassman, Jacquerod, Laepple and Schaefer — reported on their first clinical trials as follows:

> "The procaine treatment introduced by Aslan was applied in 28 cases; 22 of the patients presented diffuse or localized lesions of the central nervous system: hemiplegia (paralysis), hemiparesis (muscle weakness), spasmodic ataxia (spastic disturbance of muscle coordination), postapoplectic (following a stroke) or arteriosclerotic dementia. Each

47

patient received three 100 mg intramuscular injections a week (5 cc of the 2 per cent solution) in series of 12 injections, separated by ten-day intervals. . . . We obtained lasting results in 25 per cent of the cases, such as improvement of the subjective state, buoyancy of the psychical tonus, slight euphoria, sometimes a better physical tonus in walking and in voluntary motility, and disappearance of sphincteral incontinence."

## France

In the *Revue Française de Gérontologie* (April, 1959) Professor H. Portias of Paris, one of Prof. Aslan's early antagonists, reported on his experience with 86 aged patients whom he treated with procaine. Not being able to use placebos, he still endeavored to avoid the possibility of purely psychotherapeutical effects by telling the patients of his inexperience with the method and of his skepticism as to its efficacy.

Professor Portias' patients suffered from the usual signs of old age: arterio- and venosclerosis, chronic cough, wrinkled skin, rheumatism, and various arthritic conditions. He subjected them all to five months of procaine injections, at the customary rate of three per week; beneficial effects, except in the less severe cases of senility, could hardly be expected in so short a period. Yet in more than half of the patients the results were either "very good" (15 cases) or they were "improved" (30 cases). Twenty patients felt a bit better, 21 showed no change in their condition, and only three of the 86 showed any side effects which could be regarded as serious.

Of the 15 cases in which the results were very good, Prof. Portias noted "a physical and morale stimulation, with a disappearance of all signs of fatigue. The patients experienced a euphoric feeling. . . . disappearance of anxiety and its physical counterpart, fear, an increase in sexual power (the Russians, by the way, treat impotence with intravenous injections of procaine). . . . These results remained after the treatment was stopped." He also mentions that the cholesterol level of the blood, while initially increasing, soon falls.

Professor Portias' conclusions:

> "This is an interesting treatment, happily complementing other geriatric therapies, but no general panacea. For the time being the enthusiasm in Bucharest does not seem to be justified in all cases, even though it is true that our patients were treated for only five months. In any event it seems that the therapy is most efficacious where there exists a clear imbalance of the autonomic (involuntary) nervous system. Another very interesting technique is the combination of procaine with other substances, either simultaneously or following it."

Early last year at a meeting of the Scientific-Medical Section of the International Federation of Deportees in Vienna, Austria, the Parisian physician Dr. E.

Soladie reported on his first trials with five deportees whom he had been treating with procaine injections for several months. He made the following points:

1. The action on the skin is quite obvious; the grey appearance, so characteristic of suffering persons, is disappearing. The face takes on a natural, healthy color, the eyes become livelier. After only four months the hair shows signs of repigmentation and the zones of baldness diminish. Eczemas of different etiology have either vanished or are retrogressing.

2. None of his five patients was constantly tired as before. The irritability diminished, the sleep was better. All had reacquired a desire to work. In four of the five the appetite had improved.

3. Only in one of the five patients was there no effect with regard to sexual functions. In two, there has been a complete revival of libido (in one, after three years of enforced abstinence); in the other two, a normalization of the act.

## Germany

In a private talk with the author last year, Dr. Udo Köhler, who was one of the first to undertake clinical trials to check Prof. Aslan's results, revealed that even after giving the procaine therapy for more than two years, he is still often surprised at the results. He mentioned the case of one of his patients, a dentist, whose hobby was gardening. After the first half dozen injections, this patient asked for Dr. Köhler, who fully expected some complaint. But the dentist, a man in his middle sixties, told him: "You know how I love gardening. But in recent years I have had trouble remembering the names of the different flowers, and their Latin names escaped me completely. Now they have all of a sudden come back." And he proceeded to rattle off dozens of Latin names of the many plants he had in his garden.

More scientific evidence is presented in a case history reported by Dr. Köhler in 1958.

"Dr. G., born February 8, 1877. Symptoms of cerebral sclerosis known for 20 years, accompanied by arterial hypotension. He had been forced to retire from his profession because not only the disturbances as circumscribed by the Korsakov syndrome were apparent, but also increasing vertigo had gained in intensity to such a degree that treatment with Gerioptil (a German preparation —procaine plus vitamins) was deemed necessary. Objectively the heart muscle did not yet show any indications of a considerable myocardial damage. But many extra systoles, particularly after effort, indicated an existing hypoxia of the heart muscle. Paroxysms became so pronounced that the patient would lose his balance in changing from a lying to a sitting position.

"After only the second Gerioptil injection the patient reported a certain improvement. At the end of the first series, he was virtually without dizziness. Only in the mornings did he still observe slight dizziness—when arising too quickly. Next to the impressive eutrophic skin effect, the most remarkable sign was the return of mental capacities. The old gentleman once again participated in scientific forums, and recently—for the first time in over two years—played four-hand piano again. His partner (an internist himself) was quite astonished to find that the playing was considerably better than it had been two years previously, when the patient had abandoned the instrument because of his physical deficiencies."

In *Medizinische Monatsschrift* of May 1959, published in Frankfurt, Dr. F. Petersen, a neurologist from Halle, East Germany, reported on 137 cases of cerebral sclerosis he had treated with procaine over a period of two years (June, 1956 to June, 1958). In 13 of these cases, all symptoms of the illness disappeared; in 35, there was marked improvement; in 56, some improvement; 27 remained unchanged; and in six cases there was further deterioration. This means that in 104 (or 76 per cent) of the cases, treatment by procaine injections was to some extent successful, while other therapies offered little relief for these patients.

Doctor Petersen's figures become even more impressive if one considers only those patients suffering from a state of arteriosclerotic debility. Of the 137

| Illness | Number of patients | Cured | Marked improvement | Improvement | Unchanged | Worse |
|---|---|---|---|---|---|---|
| 1. Cerebral arteriosclerosis | 137 | 13 | 35 | 56 | 27 | 6 |
| a) Arteriosclerotic debility | 88 | 10 | 33 | 41 | 4 | — |
| b) Senile psychoses | 20 | — | — | 1 | 14 | 5 |
| c) Encephalomacie | 4 | — | — | 2 | 1 | 1 |
| d) Stroke | 8 | — | 1 | 6 | 1 | — |
| e) Parkinsonism | 9 | — | 1 | 4 | 4 | — |
| f) Cerebral arteriosclerosis and brain concussion | 4 | 2 | — | 2 | — | — |
| g) Trigeminal neuralgia with arteriosclerosis | 4 | 1 | — | — | 3 | — |
| 2. Exhaustion | 6 | 4 | 1 | 1 | — | — |
| 3. Impotence | 2 | 1 | — | 1 | — | — |
| 4. Post-encephalitic Parkinsonism | 1 | — | — | — | 1 | — |
| | 146 | 18 | 36 | 58 | 28 | 6 |

patients he treated, 88 fell into this group, and of these all but four (or 95.4 per cent) were improved: ten were considered cured, in 33, there was marked improvement, and in 41, some improvement of their complaints. Of eight stroke patients, seven showed improvement, and five out of nine patients with Parkinson's disease were also improved after the treatment. Two of the four patients suffering from encephalomalacie (softening of the brain) were also improved. But with his 20 senile psychotic patients Dr. Petersen reported almost complete failure: only one of these showed some improvement, 14 were unchanged and five actually grew worse. His statistics are shown in the table.

# Chapter 11

*Soviet research with procaine*

Not unexpectedly, medical scientists in the Soviet Union have conducted some valuable research into the properties of procaine. In a recent article in *Geriatrics,* Prof. Chauncey Leake, president-elect of the American Association for the Advancement of Science, reported:

> "When I was in Leningrad and Moscow in 1956, I discussed the trophic action of procaine with several experimental workers in the physiology research laboratories in these cities. However, I was not impressed at the time that there was anything very significant in these studies except that intramuscular injections of procaine seemed to have some benefit in experimental arthritis and perhaps in certain experimental stress conditions."

Soviet researchers, naturally, showed a lively interest in Prof. Aslan's work. Theirs has been a steady search for means of rejuvenation, but neither Voronoff (who worked outside of the Soviet Union), Bogomolets, or

Olga Lepechinskaya had any success with their ideas. The procaine therapy was given a great deal of publicity in the Soviet Union, and today there are several hospitals and old age homes where the treatment is being used, in particular at the Geriatric Institute in Kiev.

In Moscow, in 1955, Yu. F. Udalov conducted some experiments with white rats which may have significance not only in geriatrics, but in aviation and space medicine, as well as in prevention of the 'bends' suffered by underwater workers who surface too rapidly.

In Udalov's experiments, rats which had been given procaine injections in the neck and in the abdominal cavity (as well as rats which had received no injections) were 'lifted' in a baro-chamber at a speed of 67 miles per hour, to an atmosphere as thin as the atmosphere 6¾ miles above the earth. The rats were left in this 'atmosphere' for 10 minutes. Six of the 23 control rats died, but only two of the 25 procaine-treated animals. The author then found that rats given a large dose of para-aminobenzoic acid in their food the day before being subjected to these simulated high altitude conditions also had twice the survival rate of rats which had not received this vitamin. Apparently this vitamin component of procaine increases the resistance of the body to oxygen insufficiency.

A recent article in the Russian journal *Farmakologiia i Toksikologiia* (Pharmacology and Toxicology) entitled "Changes in the motor and secretory gastric functions following intravenous injection of procaine" indicates the Soviet interest in this therapy for treatment of gastric ulcers. The article ends with the following summary:

1. Intravenous injection of procaine depresses or eliminates gastric peristalsis and the periodic contractions of the stomach under starvation conditions; it also depresses the contractions of the duodenum.

2. Procaine depresses the gastric secretion, particularly in the first 3-hour period and often extending into the second period. Following injection of procaine, the latent period of gastric secretion was increased in most of the experiments. Procaine also depressed the pancreatic secretion.

3. The acid-forming function of the stomach underwent various changes under the influence of procaine, more often than not increasing.

4. Depression of the conditioned reflex of the gastric motor system after a series of intravenous injections of procaine combined with strictly controlled, fixed conditions of experiment indicated the role played by the cerebral cortex in the mechanism of action of procaine.

In addition to their interest in applications of procaine in treatment of gastric disturbances, Soviet researchers also are reporting on applications of this drug in treatment of diseases of the nervous and vascular systems, furunculosis, and bronchial asthma.

The Russian scientists seem to give Prof. Aslan full credit for originating the application of procaine in the general therapy of the condition which we know as "old age."

Among the Soviet visitors to the Bucharest Institute, Prof. M. G. Durmishyan, of the Academy of Sciences of the USSR, wrote in Prof. Aslan's guest book:

> "After my own investigation, I am able to say that the doubts I may have had before coming here have completely vanished."

Professor K. M. Bykov, head of the famous Pavlov Institute in Leningrad, expressed himself in the guest book as follows:

> "With enormous interest I saw the work of the Institute and of Prof. Aslan concerning a problem which has been studied for a long time, and [the solution of which] could revolutionize mankind. I believe that Prof. Aslan and her assistants have found a valid method of maintaining a normal state of activity of the nervous system and thus of all organs, and of prolonging the normal functions of the cellular system of the human body. The administration of procaine, which has the properties of stimulating and inhibiting, as well as the method of giving these injections, are new, original and very promising. Personally, I am

convinced of the success of Prof. Aslan's method in vitalizing aging organisms."

The intensity of the research being conducted in the USSR on the applications and mechanism of action of procaine is indicated by the fact that a recent Russian book on procaine treatment of gastric ulcers, bronchial asthma and angina pectoris cites the work of sixteen centers. This research dates back as far as 1948, and a literature reference shows that the author was familiar with the work being done in Bucharest at least a year before Western medical scientists had heard of it.

# Chapter 12

## News reaches the West

In the Western world, the dramatic story of the successes of procaine therapy reached the general public before it attracted the interest of the medical profession. In the spring of 1958, a news program of the Columbia Broadcasting System showed a three-minute documentary on Prof. Aslan's work. This short film elicited only a few requests for further information.

My own interest in this therapy had begun when, quite by accident, I accompanied a friend to hear Prof. Aslan's first lecture in Karlsruhe in 1956. As a medical writer, I was of course fascinated by what she had to report—but the antagonism, disbelief, and skepticism surrounding me were highly contagious. I wrote nothing about Prof. Aslan's claims until I received a report from my friend of the great difference between her receptions at Karlsruhe in 1956 and 1957. In December of 1958, *Coronet* published my first writing on this therapy, in an article entitled "Old drug brings new hope."

Inquiries began to pour in by the hundreds—to the magazine, to me, and even to the Institute in Bucharest. And these inquiries came not only from old men and women suffering from the very conditions described (and the families of such sufferers), but also from physicians, biochemists, and pharmacologists. The original research papers on the subject were published in translation for medical practitioners and researchers by Consultants Bureau, Inc., in March of 1959.

At about this same time, the London *Daily Mail* published a series of five articles by a woman reporter who had gone to Bucharest to learn at first hand the value of this therapy. Her highly sensational articles aroused such great hopes and expectations on the part of the aged population of London, as evidenced by the thousands of letters received by the paper, that the series was concluded with a note from the paper's science editor, quoting an unnamed British doctor:

"Old people taken out of lonely or unhappy or disease-prone backgrounds and given expert attention and encouragement in cheerful surroundings often take on a new lease of life.

"Much of the evidence at Bucharest, say visiting experts, may be based on hearsay—patients' ages, previous conditions, and so on.

"The same drug used by Prof. Anna Aslan has been tested in this country 'without any startling results'."

To which the science editor added:

"There is no scientific reason known to me why it should cause 'rejuvenation', one specialist told me. On the other hand reasons sometimes come after results. Knowledge of digitalis came after a general practitioner found that an infusion of foxglove did his heart patients good."

On March 14, 1959, a few weeks after the appearance of these articles, the well-known British medical journal *The Lancet* discussed the procaine therapy in an editorial. It said:

"The cause of the decline of vigor in mammals with age is unknown. There is no a priori reason why procaine, or many other uninvestigated substances, should not slow or even reverse this decline, and a substance which did so would quite possibly produce just the kind of non-specific benefit in a number of disorders which Aslan describes. The regrowth of pigmented hair in a man who claimed to be 110 years old, which Aslan reports, would in any case, like the validity of the age record, excite curiosity. But it is curiosity rather than enthusiasm that Aslan's treatment of her results is likely to excite. Her suggestion that procaine acts by in vivo (inside the body) conversion to p-aminobenzoic acid, and that this exercises a "trophic" action on the nervous system, does not carry instant conviction; moreover, such an influence (or, in fact, any specific benefit from a drug administered to geriatric patients to control "aging") could be shown convincingly in one way only—by a double-blind trial in alternate matched cases, with subsequent comparison of objective signs and survival curves. All the published evidence so far depends on scientific medicine's chief methodological enemy—the testimonial use of case histories—and Aslan's treatment of these will depress those who know how often medical investigators have misled themselves in this way. There are very few old people who do not respond to rest, change, good hospital food, and, above all, raised morale—whether accompanied by injections of procaine or not.

"This is not to say that the work of the Bucharest team is to be dismissed (they have evidently improved their patients in some way, if only by suggestion); and the desire to do something radical about old age is a creditable contrast to fatalism about the effects of age. If these workers were right, the findings would be important. But the facts can be established only by properly controlled trials."

In the United States, publication of the translated reports from Bucharest was not met with indignation, as was Prof. Aslan's first lecture at Karlsruhe. But the reception was, to put it most mildly, unenthusiastic. This lack of enthusiasm was not, as some have suggested, because the new therapy had been developed in an Iron Curtain country. It was because the methods of analyzing the results of the therapy fell so far short of

American research standards. American researchers feared that the lady from Rumania who was making such extravagant claims for procaine therapy had fooled herself, as well as her patients, into believing that they felt and looked so much better. (Later on we shall see why even the substantiating reports from other researchers and practitioners in other European countries are not totally convincing to the Western scientist.)

But there was also a sober recognition that the reports by the Bucharest and German doctors should not be ignored. Prof. Leake (Ohio State University), who reviewed the translations of the Karlsruhe papers in the journal *Geriatrics* in October, 1959, said:

"In general, it would seem that the reports by Anna Aslan and her associates are interesting enough for further exploration. It would be hoped that she and her associates would publish more detailed case histories, together with a more complete statistical survey of the large number of cases which they must have accumulated by now. It would seem that careful and direct experimental studies on small animals should tell readily whether or not repeated procaine hydrochloride injections intramuscularly can delay the aging process, prolong life, and generally interfere with aging. It might be wise for results of studies of this sort to be well publicized before extensive premature clinical use of procaine hydrochloride in slow-

ing the aging process. On the other hand, the safety of
the drug indicates that cautious and well controlled clinical studies might yield results that would tell definitely
whether or not any further use of the drug for these purposes is justified."

Little can be added to what has been so calmly and
objectively pointed out by the *Lancet* and *Geriatrics*.
Both medical journals have emphasized the necessity
that the work Prof. Aslan began at the Institute of
Geriatrics in Bucharest be subjected to the most careful investigation on as large a scale as possible, with
the use of the most modern equipment available and
the most rigid control standards devisable.

# Chapter 13

## The Parhon Institute of Geriatrics

It is no exaggeration to classify most homes for the aged as brick and mortar limbos, wherein old people in both physical and mental states of dissolution continue their slow degeneration amid the organized monotony of white bedsheets and hushed voices, broken only by the seventh day visit of an impatient relative. Rehabilitation on this basis is an impossibility, and the most that can be hoped for is constant sedation which will relieve the aches and pains, the discomfort and discontent, the feeling of having been forgotten and the frustration of being able to remember.

Each of the patients in the Bucharest old age home was, at one time, debilitated, plagued by sickness, and steadily growing weaker in mind and body. Today, not one of the patients who has received the therapy is bedridden! That in itself is a remarkable achievement. However, the procaine-treated patients not only walk about unaided; they also function as alert, thinking, industrious men and women who are old without being aged. Almost every one of them has a task: some work in the kitchen, others in the garden; a few are busy carpet weaving, others make handicrafts, and some help in the library. And, most significantly, many of them even attend courses in French. The procaine therapy has helped them to remain mature while they grow older.

The research upon which the whole therapy is based was begun in the Institute in May of 1951. Twenty-five patients, ages 60 to 92 (all of whom suffered from serious degenerative diseases such as extrapyramidal disturbances, hypertension, degenerative joint disease, rheumatism, cirrhosis, etc.), were treated with procaine. After some initial uncertainty, Prof. Aslan arrived at what she believes is the optimal dose and strength for the procaine: intramuscular injections of 5 cc of a 2 per cent solution, three times weekly for four weeks, were administered. A ten-day interval without further inoculation followed, then a new series of twelve injections, another interval, and so on.

Before treatment was started, possible allergic reaction was tested for with an injection under the skin of 1 cc of procaine. Few people were found to be sensitive to this substance—in Bucharest only a handful among the thousands of patients (four out of 4,800 tested). It can reasonably be expected, however, that a higher allergy rate may be found in the West, where sensitization generally is greater and allergic disorders affect more people than in Eastern Europe.

Seventeen of the 25 old men and women in Prof. Aslan's original group were still alive eight years later, in spite of the fact that the disorders from which they had been suffering (as stated above) would ordinarily have claimed the lives of almost all of them. There was not a single death during the first two years of treatment. In 1954, there was one death due to a spinal accident; in 1955, two deaths occurred, due to chronic bronchitis and myocarditis (an inflammation of the muscular part of the heart wall); in 1956, one death due to a pseudobulbar syndrome, from which this patient had suffered for six years. Five patients died in 1959 — three in a flu epidemic, one due to arteriosclerosis, and one due to an accident.

The number of patients receiving procaine therapy was slowly increased after the successful results with the first group tested. Today, 70 of the patients in the old age home receive procaine injections, 40 other inmates being used as control groups. The table indicates a portion of the impressive results obtained.

| Year | No. of patients | Able to work | | Unable to work | | | | | | Mortality | | |
|---|---|---|---|---|---|---|---|---|---|---|---|---|
| | | | | Capable of self-care | | Incapable of self-care | | Total | | | | |
| | | No. | % | No. | % | No. | % | No. | % | No. | % | Age |
| 1951 | 25 | 12 | 48 | 13 | 52 | — | — | 13 | 52 | — | — | — |
| 1952 | 29 | 20 | 68 | 9 | 32 | — | — | 9 | 32 | — | — | — |
| 1953 | 34 | 24 | 70 | 10 | 30 | — | — | 10 | 30 | 1 | 2.9 | 86 |
| 1954 | 45 | 31 | 71 | 14 | 29 | — | — | 14 | 29 | 2 | 4.4 | 74;78 |
| 1955 | 45 | 36 | 80 | 9 | 20 | — | — | 9 | 20 | 1 | 2.2 | 89 |
| 1956 | 66 | 55 | 84 | 11 | 16 | — | — | 11 | 16 | 2 | 3.0 | 86;71 |
| 1957 | 72 | 63 | 87 | 9 | 13 | — | — | 9 | 13 | 3 | 4.2 | 79;98;74 |
| 1958 | 72 | 65 | 90.3 | 3 | 4.2 | 4[1] | 5.5 | 7 | 9.7 | 2 | 2.7 | 93;72 (accident) |
| 1959 (before flu epidemic) | 70 | 64 | 91.4 | 3 | 4.3 | 3[1] | 4.3 | 6 | 8.6 | — | — | — |
| 1959 (first nine months including flu epidemic) | 67 | 60 | 89.7 | 4 | 5.9 | 3[1] | 4.4 | 7 | 10.3 | 5[2] | 7.4 | 77;84;92; 96;104 |
| Mean | 52[3] | | | Increase in capacity to work: 83.7% | | | | | | | 3.8[4] | 84[5] |

[1]Recently hospitalized cases
[2]Three deaths due to flu, one due to arteriosclerosis, one due to accident
[3]Median number of patients
[4]Average mortality
[5]Median age

69

The mere prolongation of life was of minor concern to the Bucharest doctors. (Indeed, most of the patients there had already achieved an enviable record of longevity.) Instead, they were interested in developing a method of the preservation and restoration of vitality. Thus, the figures indicating the improved capacity of the procaine patients to do some work and to care for themselves are much more significant than the dramatic figures on mortality. These exciting statistics are shown graphically in the photo section (pp. 81-96).

When I first walked among the patients at the Bucharest Institute, I was struck by the fact that some seemed to be merely well preserved, active people for their apparent age, while others had a strikingly *vital* appearance (I hesitate to use the word 'youthful' lest a reader misinterpret and think I am implying that some of the patients had grown younger). *All of the patients, including those who have been receiving procaine more than 100 times a year for eight years, remain old people.* But those who have been on procaine therapy for some time have a look of vitality that one might characterize as "80-years-young," or even "112-years-young" in the case of Parseh Margosian.

Until the flu epidemic in 1958/59, the mortality of

the patients treated with procaine was 3.2 per cent (and the median age 82 years), that of the people treated with vitamins or gland extracts was 16 per cent, and of those who did not receive any supportive therapy, 27 per cent. Since the patients themselves do not know what treatment they are receiving, the great difference in the mortality of these groups contradicts those critics who ascribe the success of the procaine therapy to some suggestive effect.

The Institute of Geriatrics was founded in 1951 to study the problems of gerontology and geriatrics, as part of the research plan of the Academy of the Rumanian People's Republic, thus constituting a portion of the State Scientific Program. Therefore, it is important to note that Prof. Aslan was not, as were most of her predecessors in the application of procaine, working on an individual basis with the goal of personal symptomatic relief. One hundred and eighty-nine patients are being treated today in the Institute, as part of a large government-sponsored research program.

As a result of the initial success with the procaine program, the Institute has become physically larger and scientifically a more important place. It is now composed of five departments:

1. The nursing home, with 110 beds, where aged

people are cared for under the conditions cited previously (70 patients receive procaine and 40 are used as controls). Long-term treatment is provided here.

2. The clinic, with 80 beds, for the treatment of bedridden aged patients as well as others afflicted with certain diseases which respond to procaine treatments. In this section, the emphasis is more on short-term treatment.

3. The out-patient department, where procaine treatments are administered daily to 600 to 700 persons, some for therapeutic and other for prophylactic purposes. (Ninety per cent of all patients are treated with procaine, while the other 10 per cent are used as a control.)

4. The laboratories: some for animal experiments, others for clinical, physiological, biochemical, hematological, pharmaceutical and roentgenological research. Each of these laboratories is under the direction of a specialist, some of whom are visiting experts from Rumanian Universities.

5. The department of social hygiene, which is mostly concerned with the sociological problems of old age and relates national statistics to the work of the Institute.

The fact that the Institute's research functions have become increasingly important is reflected in a contemplated change of name from the Institute of Geriatrics to the Institute of Gerontology. Because the Institute will continue to treat, as well as to investigate, it is distinguished from almost all other existing centers in this field, most of which are devoted exclusively either to the care of the aged or to research into their problems.

# Chapter 14

## The procaine therapy given at the Institute

The hospital division of the Institute has slowly expanded as it became apparent that a number of diseases which required hospitalization could also be treated through long-term procaine injections. As a matter of fact, increasing numbers of children are brought to the Institute's hospital or to its out-patient clinic for the treatment of such disorders as skin diseases, osteoporosis (enlargement of bone marrow), bronchial asthma and rheumatism.

There are also rare cases of alopecia (or baldness) and vitiligo (the loss of pigmentation in skin or hair, making it appear white, which can occur in children and young adults as well as people of more advanced age). The great majority of the patients being treated with procaine today are out-patients. They are under constant observation, but of course cannot be controlled as carefully as the inmates.

The treatments for patients of the Institute, however, have become standardized.

The procaine used has been modified over the years. While it originally had a pH (hydrogen exponent) between 4.2 and 5, this has now been reduced to 3.7 in the Rumanian preparation called Gerovital or "$H_3$," produced according to Prof. Aslan's formula. (A substance at a pH of 7 is neutral — neither acid nor alkaline, but below 7 it becomes acid.) Experiments proved that if the pH of procaine is augmented, its anesthetic properties are increased; conversely, if the pH is reduced, it loses the anesthetic properties, while the action on the sympathetic and parasympathetic nervous system is preserved or even increased. The lowered pH may have something to do with the fact that, in Bucharest, side effects have almost never been encountered, and that allergic sensitivity to the stabilized substance is almost nonexistent.

Intramuscular injections of procaine are the rule in this treatment, but there are a few exceptions. In cases of asthma, gastric or duodenal ulcers, and vascular spasms, intravenous injections are considered more effective, and in arteritis as well as certain arthropathies, intra-arterial ones are indicated. The rhythm of injection is also slightly changed: in vascular spasms, 5 cc is given twice daily; in ulcers, 5 cc is given once daily at a very slow rate. The usual series of twelve injections with a rest period of seven to ten days is constantly maintained, however.

The 5 cc dose of procaine is reduced only where body weight of the patient is abnormally low; in these cases 3 or 4 cc are recommended. Children, as a rule, tolerate the 5 cc dose well. In the prophylactic treatment of old age, dealing with people between 45 and 60, the cycle of twelve injections per month is observed, but one or two series per year suffice, whereas in the geriatric clinic, procaine is given as a maintenance dose. The prophylactic treatment is of course in the purely experimental stage, and it will be many years before any significant statistical data can be obtained.

As further proof of the fact that one does not develop a tolerance to procaine, it is important to note that even among the survivors of the original group which Prof. Aslan has worked with and treated for more than eight years (they received more than 1,000 injections), higher doses are not needed in order to achieve the desired effects. Nor has a sensibilization for procaine been noticed.

There is only one positive contraindication of the

procaine therapy (with the exception of the allergy, of course), and this is the simultaneous administration of sulfa drugs. Here, procaine acts as an inhibitor. If plain procaine is used for the treatment, the patient should, as a precautionary measure, rest for about half an hour after the intramuscular injection. Longer rest periods are indicated in cases of intravenous injections. Procaine has a dilating effect on the blood vessels, causing a lowered blood pressure, which in turn can produce untoward side effects. When a buffered form of procaine is used, however, there is no need for rest after the intramuscular injection.

Other procaine preparations of slightly different compositions which have been developed in other countries are: "Gerioptil", "Gerioptil plus $H_3$" (the latter being used for oral therapy)—combinations of dimethylaminoethanol and p-aminobenzoic acid, folic acid, several vitamins, unsaturated fatty acids and other substances), "Prokopin-G" and "Ruticain" (a combination of procaine, rutin, and Vitamin $B_{12}$) in Western Germany; "Procaine Vifor" in Switzerland; "Novaine" and "Procaine Minerva" in Greece; "Trophomorina-$H_3$" in Spain; "Compensol" and "Gericain" in Argentina; "Gerontex-$H_3$" in Brazil; and in Ecuador, "Juventocain."

For all these pharmaceutical products, the therapy

prescribed follows closely the plan developed by Prof. Aslan. All these preparations must be administered by injection, with the exception of "Gerioptil plus $H_3$," which is available in capsule form. But the efficacy of the latter as far as older people are concerned is doubtful. Older men and women lose the ability to absorb vitamins through the digestive system, making it difficult to arrive at a satisfactory dosage. In order to solve this problem, old people are first given twelve injections of "Gerioptil," after which they are put on capsules until another series of injections is deemed necessary.

In the United States, procaine under many trade names is available from various pharmaceutical companies.

It is important to clear up the confusion that exists in the minds of many doctors as to whether the procaine available from pharmaceutical companies in the United States could be expected to be as effective as the "Gerovital" preparation. Prof. Aslan has never claimed any mystery ingredient in her formula, and has always identified procaine as the active ingredient. Her original research of the "25 cases" and the "100 cases" was conducted with a solution of procaine at a pH of 4.2. When she lectured recently in London to the Medical Society for Care of the Elderly, she stated, in response to a direct question, that no trace elements are given along with the procaine in the Rumanian preparation.

Understandably, the Rumanian chemical export company likes to sell its product in other countries, and the precise formula has not been published. But *everything Prof. Aslan has ever written or said indicates that procaine is the active ingredient responsible for the results which she reports.* Thus, procaine at the required pH (some procaine products on the United States market are already at exactly the prescribed pH) should have the same effectiveness as Gerovital.

Certainly no one should pay exhorbitant prices for Gerovital vials reputedly "smuggled" into this country, for the cost of enough procaine for a single injection is so low that the price to the patient should not be above a doctor's usual fee for a routine injection.

Whether the entire procaine therapy may be simplified some day to the point where it will be possible to develop an effective oral preparation cannot be predicted. Much more research will be necessary before the theoretical possibility of such a preparation becomes an actuality.

# Chapter 15

## *A rehabilitator for the aged?*

As we have mentioned previously, procaine acts on many parts of the body. Clinical observations lend credence to the findings that procaine affects almost all those organs, glands, and functions that are particularly deficient in old age. It is important to bear in mind that most of these conditions are of a chronic nature and require long-term treatment. Many of the failures with certain applications of the procaine therapy may stem from the fact that the treatment was not applied for a sufficiently extended period of time.

The data in the table cover a significantly large number of patients at the Institute who were treated during the years from 1952 through 1958, for a variety of diseases. Again, the statistics on improvement are much more convincing than those showing a lower mortality rate as compared to patients not treated with procaine, as we shall see in Chapter 26.

"Professor Aslan showed a number of black-and-white and colored slides of these patients to demonstrate changes in their external appearance following treatment. But, as no two photographs of the same patient had been taken under the same conditions of lighting and background, these were not entirely satisfactory."

*British Medical Journal,*
Nov. 28, 1959.

"Before and after" pictures are convincing to the medical scientist only when it is obviously the patient who has changed, and not merely the angle of photography, the lighting, or the type of film used.

The four pictures of Maria Tabarcea shown in these pages are excellent documentation of her case history. The improvement in the condition of Tanasalu Mircea is obvious, in spite of the difference in lighting. But as medical documentation, the other photographs of patients are "not entirely satisfactory."

Professor Anna Aslan, M.D.

*Time* labeled her loosely "Dr. Faust in skirts."

The London *Daily Mail* called her "The Live-Longer Professor."

*Geriatrics* said her reports are "interesting enough for further exploration."

The C. I. Parhon Institute of Geriatrics is housed in this building, which is considerably larger than the picture indicates. The laboratories in the basement have recently been expanded and their equipment improved.

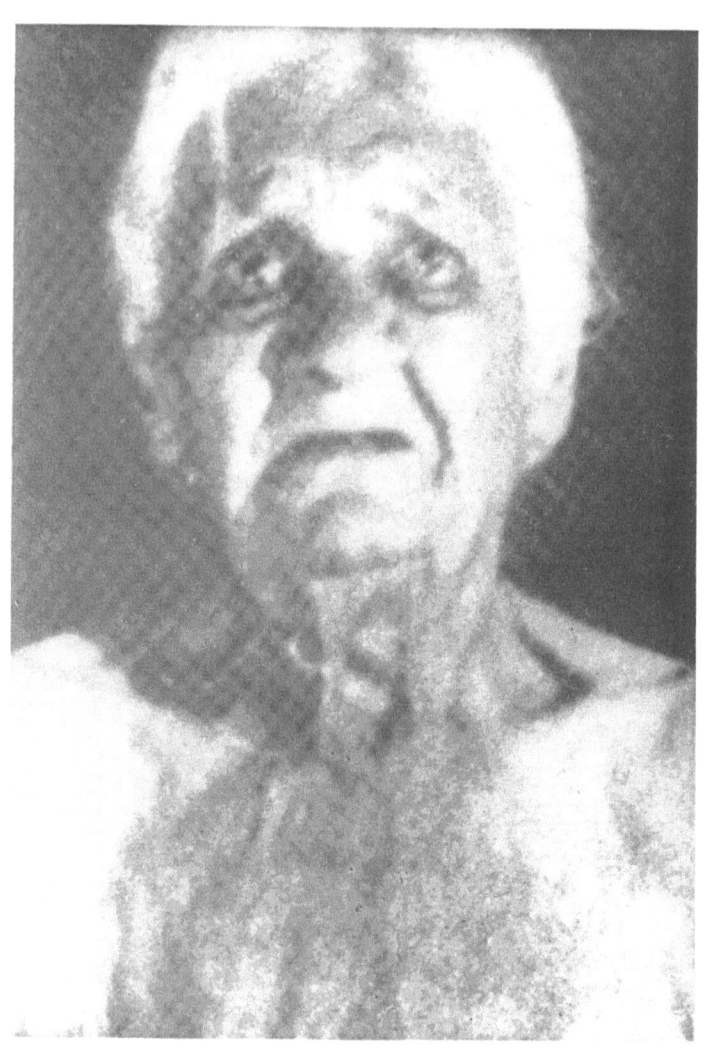

A woman of 113 years, when first seen at the Institute.

The same woman, after one year's treatment with procaine injections. (The patient has since died.)

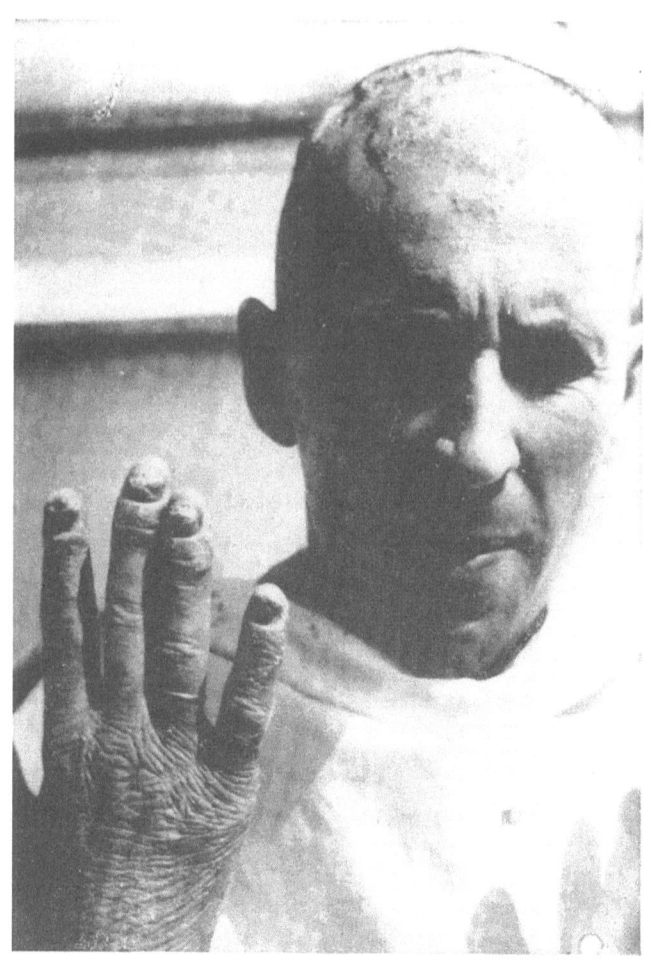

Tanasalu Mircea as he appeared when coming to the Institute. He was only 42 years old, but suffered from arthralgia, and a severe case of psoriasis, of 17 years standing. He had been treated with other drugs for five years.

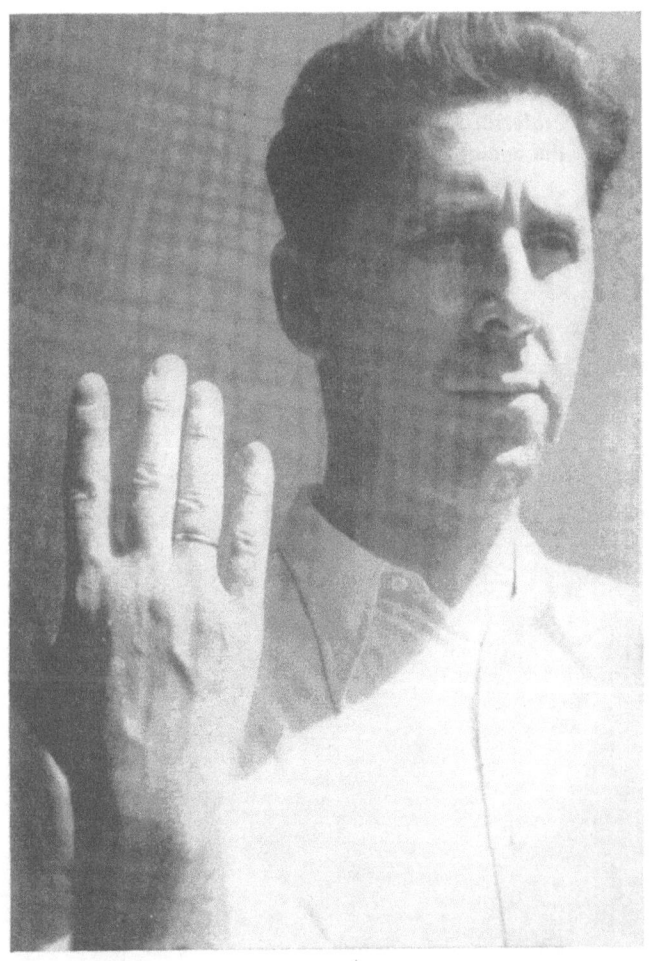

The same man when he was discharged as cured
only one year later.

Professor S.P. before beginning the treatment at the age of 92.

**Prof. S.P. shortly before his death (due to influenza) five years later.**

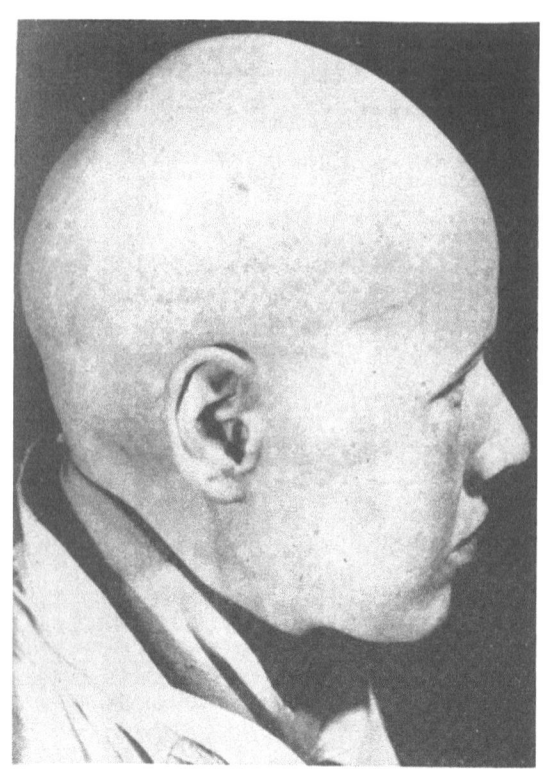

Maria Tabarcea was suf-
fering from complete
baldheadedness (alope-
cia) when the picture at
left was taken ... Then
procaine treatment was
begun.

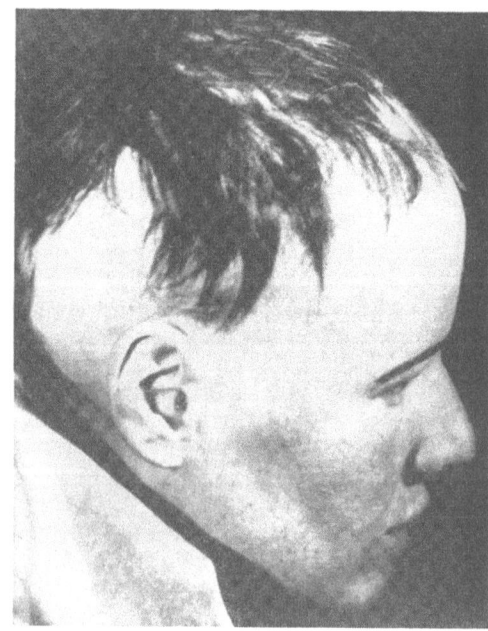

Right: One-half year
later.

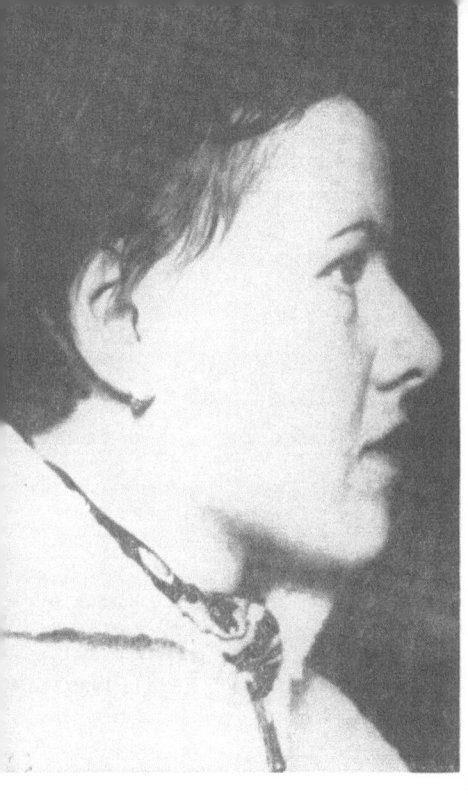

One-and-one-half years later, Maria was completely cured.

Because alopecia sometimes undergoes a spontaneous cure, even these excellent documenting photos are not absolute proof that procaine caused the cure, but they are valid and convincing evidence.

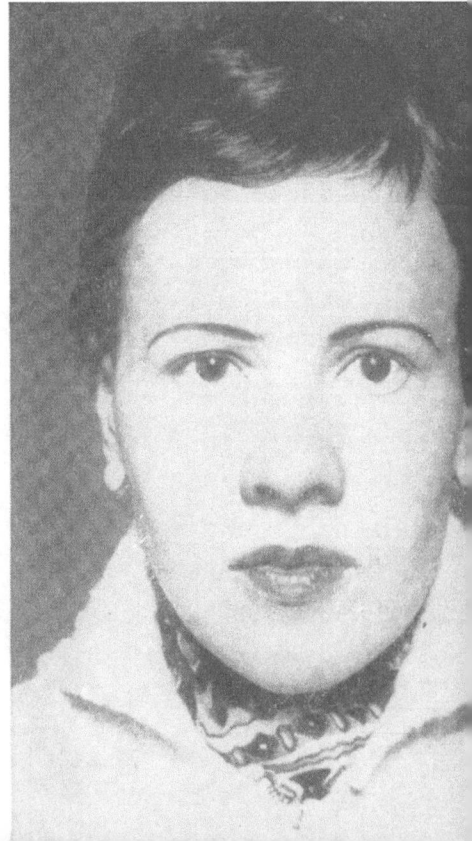

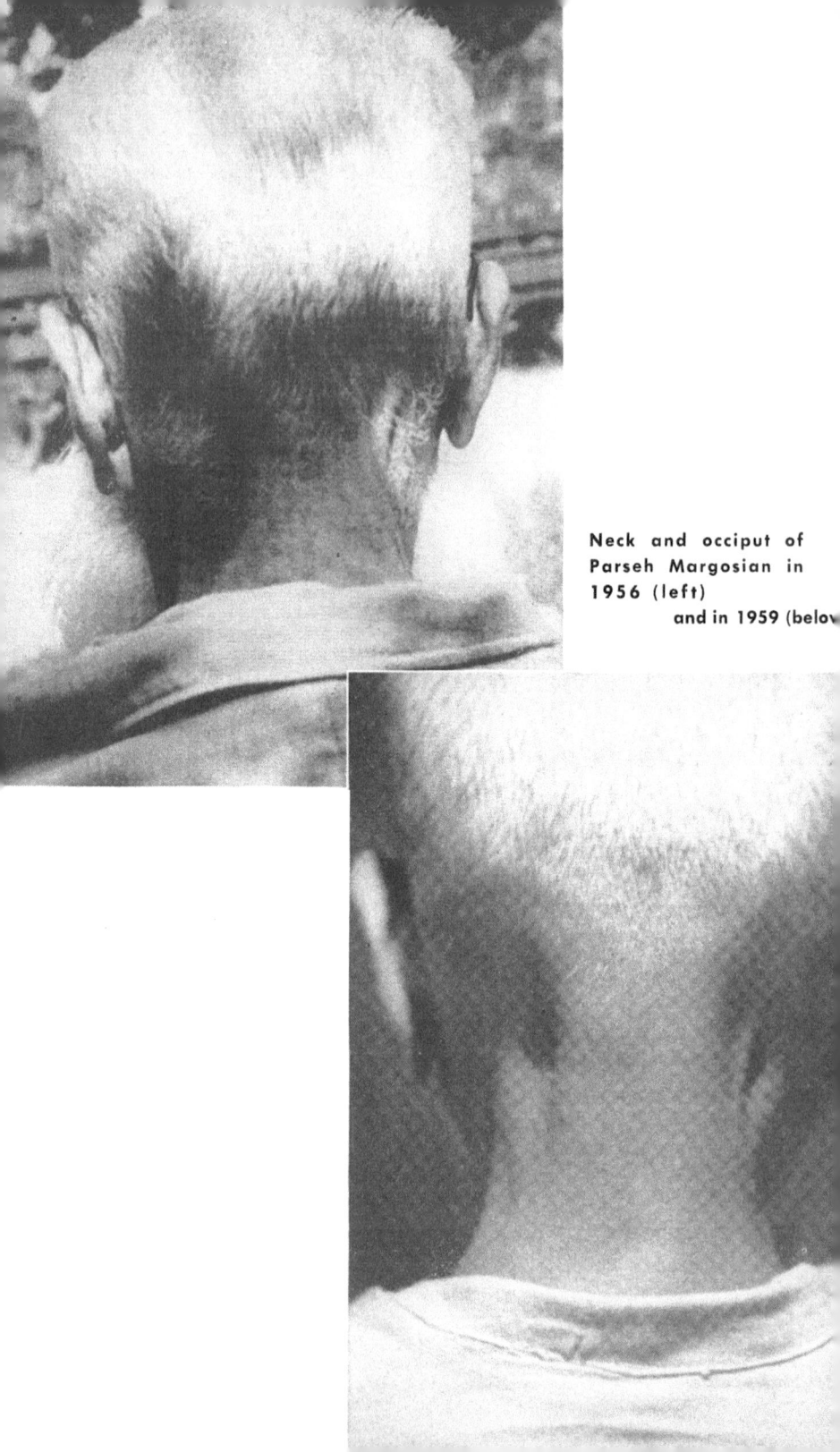

Neck and occiput of
Parseh Margosian in
1956 (left)
    and in 1959 (below)

Mr. Margosian, at 112 years, is the oldest inhabitant of the old age home, and is still alert and active. Of course, if his age records are correct, he is obviously an exceptional man, who lived to be 106 years old without benefit of procaine injections.

Octogenarians working in the carpet weavery at the Institute.

Some of the handicraft objects made by the inmates of the old age home of the Institute.

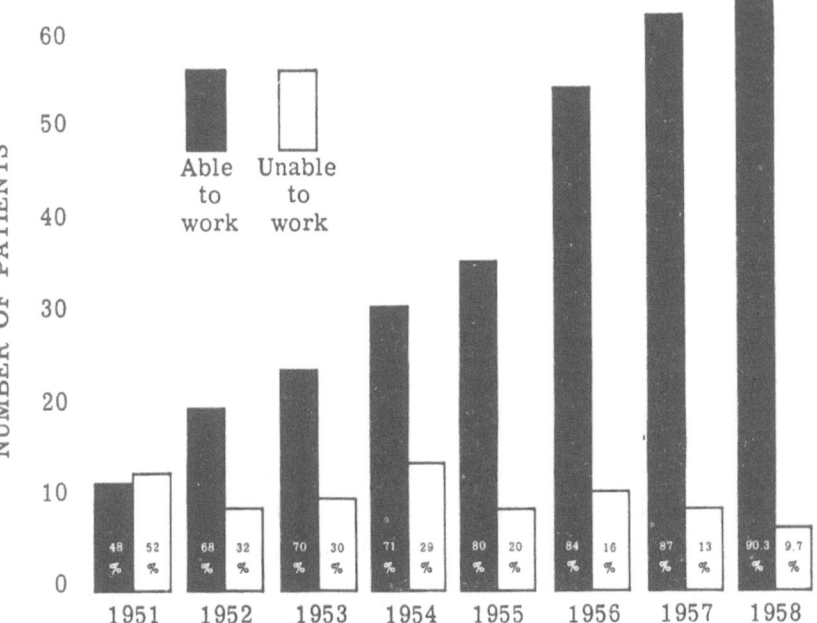

NUMBER OF PATIENTS

| 1951 | 1952 | 1953 | 1954 | 1955 | 1956 | 1957 | 1958 |
|------|------|------|------|------|------|------|------|
| 48% 52% | 68% 32% | 70% 30% | 71% 29% | 80% 20% | 84% 16% | 87% 13% | 90.3% 9.7% |

Able to work   Unable to work

**This chart shows the dramatic improvement in the percentage of patients able to do some work as the procaine treatment program has been continued and expanded.**

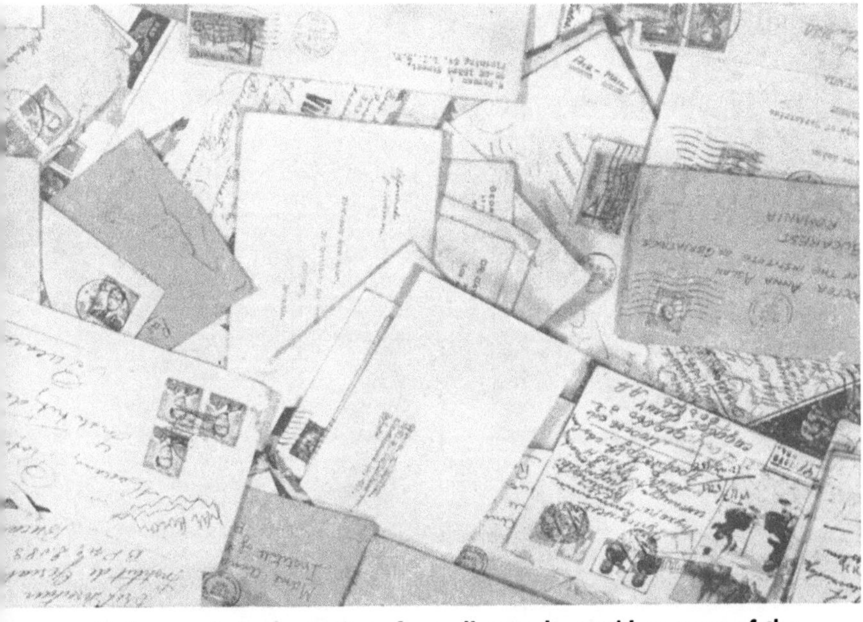

**Mail pours in to the Institute from all over the world as news of the**

A trained medical photographer recently joined the staff of the Institute of Geriatrics, as part of the expansion and improvement of its research facilities.

Documentation by medically acceptable "before and after" photographs would help to dispel the skepticism of British and American doctors, who fear Prof. Aslan has allowed her enthusiasm to destroy her objectivity.

Let us hope that such documentation may be available shortly.

| Disease group | No. | Improved | | | Unchanged | | | Deaths | | |
|---|---|---|---|---|---|---|---|---|---|---|
| | | No. | % of disease group | % of total | No. | % of disease group | % of total | No. | % of disease group | % of total |
| 1. Nervous system | 300 | 245 | 81.7 | 28.0 | 43 | 14.3 | 4.9 | 12 | 4.0 | 1.4 |
| 2. Locomotor | 226 | 217 | 96.0 | 24.0 | 9 | 4.0 | 1.0 | — | — | — |
| 3. Cardiovascular system | 297 | 269 | 90.6 | 30.6 | 16 | 5.4 | 1.8 | 12 | 4.0 | 1.4 |
| 4. Skin and hair | 37 | 34 | 91.9 | 4.0 | 3 | 8.1 | 0.3 | — | — | — |
| 5. Gastric ulcers | 15 | 15 | 100 | 1.7 | — | — | — | — | — | — |
| Total | 875 | 780 | — | 89.1 | 71 | — | 8.2 | 24 | — | 2.7 |

Patients who were not treated with procaine:

| | No. | Deaths | % |
|---|---|---|---|
| | 495 | 54 | 10.3 |

What kind of diseases yield best to the procaine therapy, quite apart from the general improvement noticed in the patients? The statistics in this table indicate that diseases affecting the skin and hair, as well as the wide circle of ailments connected with the central nervous system, are most responsive. Among the former we find eczemas of different etiology, alopecia, herpes zoster (shingles), psoriasis, vitiligo, ichthyosis and scleroderma, to name but a few where some experience has been gained. Among the latter gratifying results have been achieved in post-apoplectic situations, paralytic states, neuralgia and neuritis, Parkinsonism (as already mentioned), Burger's disease and multiple sclerosis. The application of procaine therapy is also fruitful in degenerative joint and bone diseases, such as arthritis and the different arthroses, osteoporosis, and Bechterew's disease.

Before we discuss the various theories as to how procaine acts on the various parts of the body to affect the diseases which attack these parts, let us first examine in more detail the extent to which the body and its diseases are affected by procaine therapy.

# Chapter 16

## *Diseases of the nervous system*

The effects of procaine on the nervous system appear relatively fast, and are clearly noticeable. The disorientation and confusion which is so often characteristic of old people disappears, and memory, perception and ability to concentrate are renewed. In younger people under treatment these mental functions seem improved as compared to their capacities before treatment with procaine. Shortly after treatment is initiated most old people show an increased desire to live: they display better moods, their eyes become increasingly bright, and their hearing (and occasionally their vision) improves. This indicates that procaine injections produce an immediately increased response to stimuli. Perhaps the ganglions of the diencephalon (interbrain) are affected, leading to an improvement in walking ability, to a better mobility of the fingers, and to a decrease in so-called extrapyramidal rigidity (characteristic of

Parkinson's disease and other illnesses in which muscles stiffen and the expression of the face becomes masklike).

According to Prof. Aslan, procaine has a neurotrophic action (supplies the nerves with nourishment) not only upon the central nervous system, but also upon the peripheral nervous system, which would explain its effect upon a great variety of diseases. Italian researchers have found that procaine exerts a direct action on the brain; after intravenous injection of procaine in animals, they found the largest quantity of it in the brain. Russian physiologists have also confirmed the direct action of procaine on the nervous system.

Studies at the Institute have shown that it is within the highest age brackets that neurological diseases are by far most frequent, whereas cardiovascular and rheumatic diseases are the main problems in advanced middle age. In view of these findings it is particularly important to note that procaine also affects the peripheral nervous system, as evidenced by a cessation of neuralgia and a decrease in neuritis, both of which are very painful conditions frequently present in elderly people.

Conditioned reflexes also are improved under procaine therapy. After prolonged treatment older people are able to fix their reflexes after only three associations, which is the rate generally observed in young people and adults (usually, in older people from nine to twelve associations are required). No similar improvement was noted in the elderly patients who were treated with any other substances, with the exception of thyroid extract, which has a slightly stimulating effect upon the central nervous system.

Thus far, there is not enough experience to reveal the whole potential of procaine therapy in special neurological cases, nor has the method been sufficiently refined to ensure the quickest possible recovery of the patient. Prof. Aslan's technique is nonspecific (stimulating the entire organism, and thereby indirectly influencing the ailment instead of attacking the specific ailment directly).

**Multiple sclerosis**

A case in point is multiple sclerosis, a disease still considered incurable. Prof. Aslan does not claim that she is able to cure it; in the first place, she has not had enough cases, nor have those patients been observed for a long enough period. Still, the fact that she has been able to achieve considerable improvement in her multiple sclerosis patients (with many series of procaine injections) is undeniable. Again, the use of procaine in multiple sclerosis is not quite new, and this fact underscores Prof. Aslan's repeated statement that hers is not a discovery, but a rediscovery.

We should also note that there are periods of spontaneous remission of this disease, usually followed by a worsening of the condition.

In 1950, the noted West Berlin surgeon Erwin Gohrbandt reported dramatic improvement as well as cures of multiple sclerosis with procaine injections into the sympathetic trunk, in particular into the stellate ganglion and the solar plexus. By 1951 he had treated 87 persons, and in a few advanced bedridden cases freedom from all symptoms had lasted for three years, which is considerably longer than the usual periods of retrogression of this disease. He also noted, however, that the stage of the illness at which procaine treatment is begun seems to determine the possibility of success.

**Parkinson's disease**

As early as 1919, Dr. G. Liljestrand described highly promising results with procaine in the treatment of Parkinson's disease, but no subsequent reports were made. It was not until more than 35 years later, when Prof. Aslan reported her success in reversing the Parkinson syndrome (rigidity of muscles, tremor of the arms and hands, loss of associated and automatic movements, masklike facial expression), that the earlier paper was remembered. In the years preceding Prof. Aslan's rediscovery, no one had paid any attention to the 1919 paper.

## Postapoplectic conditions

Another rather important application is the intravenous injection of procaine in cases of apoplectic coma (unconsciousness after cerebral stroke). The patient is usually brought back to consciousness quite rapidly and this status is maintained from 30 minutes to several hours, depending upon the severity of the attack. In any event, there is sufficient time in which to feed the patient and thus banish the danger of aspiration pneumonia (caused by foreign bodies being drawn into the lungs while the swallowing center is not functioning). This latter complication cannot be fought even with antibiotics.

## Loss of hearing

A very interesting experiment was conducted by Dr. P. Braunsteiner, of Rheine, Westphalia, Germany. This physician confined himself to the observation of a group of elderly hard-of-hearing patients. He chose 35 people over 55 years old, and subjected them to several series of procaine injections. All reported feeling generally better, and gave evidence of increased auditory acuity. Audiometric measurements showed that in 13 of the patients the improvement in acuity was strictly subjective—they were all over 70, and probably had already suffered irreversible damage to the nerve cells. In the other 22, the improvement could easily be verified by the usual methods and the audiogram showed an increase in acuity of 15 to 30 decibels.

"This audiometric proof is clinically very interesting," Dr. Braunsteiner reports, "but we feel that the subjective improvement is of larger practical importance. This subjective improvement without exception is greater than compatible with the audiometric results. The reason must be looked for in the effect of Gerioptil on the general condition of aging people. All patients feel more vigorous, mentally as well as bodily, their memory is improved and the general interest in their surroundings has increased. Consequently the power of concentration and receptivity has also increased, and with it the perceptive faculty of the ears. An examination of the acuity for voice showed an average increase in hearing power for conversational speech from 2 to 3 meters, and quite frequently even 4 to 5 meters. The perceptivity for whispering was only slightly improved."

# Chapter 17

## *Diseases of the muscles and joints*

The aches and pains which plague so many of the aged are usually connected with either the muscles (particularly the legs and hands) or the joints. Procaine has been found to be highly effective in relief of such pain by many workers, and its therapeutic effects on these conditions were noted long before Prof. Aslan began her work.

### Rheumatism and arthritis

As we mentioned earlier, Gustav Spiess was the first to point to procaine as an antirheumatic agent. However, he did not pursue his original observations, and they were soon forgotten. It was not until the nineteen-forties that Dr. M. G. Good of the Charterhouse Rheumatism Clinic in London championed procaine treatment for his patients. In this country, Dr. David J. Graubard, of New York City, devoted himself to its application, and wrote extensively about it.

The effect of procaine on the joints was the point of departure for the present therapy. Successes were perhaps less pronounced than in other areas for which procaine therapy was indicated, although, in the words of Dr. Good, "the results of procaine treatment (in rheumatic conditions) would be hard to match or surpass by other methods of therapy." Frequently, the mobility of the joints is restored, the pain alleviated, contracture of muscles decreased, and the muscle power strengthened. Prof. Aslan reports that X-ray pictures sometimes indicate remineralization of the bones, which may be the reason why fractures, seemingly, heal faster with procaine therapy.

In a lecture before the Congress of Internal Medicine in Paris, in September, 1950, Dr. Good described his successes in treating muscular rheumatism as well as arthritis with intramuscular injections of procaine (a technique little used heretofore). In rheumatism the pains are concentrated in the so-called "myalgic spots," which are also very sensitive to pressure. His report covered 80 cases, and

> ". . . . the therapeutical results were brilliant, often dramatic and wonderful. A permanent cure can be forecast with great probability, if not positive safety. In clinical medicine there are few therapeutic measures which allow more impressive and dramatic successes: a patient,

for instance, who is suffering from an acute attack of lumbago . . . . and complains about unbearable pains, can be freed from his complaints, as with a wand, within a few minutes through the injection of 5 - 10 cc of novocain. This method was tried out in the British Army during the war in many cases, and found to be very successful."

Doctor Good finally makes the following observation, which is quite significant in view of Professor Aslan's later findings:

"Another favorable side effect of the novocain therapy also must be mentioned, the improvement in the general well-being, and quite often a better mood of the patient. Patients who have been ailing and plagued by pains for years often seem quite depressed; they have dull eyes and are rather egocentric. They do not take any interest in people around them; their main occupation consists in an endless reciting of their ills and in tyrannizing other people. After a few weeks of continued novocain therapy the favorable change in the state of the chronically ill can readily be seen: Their glance is no longer directed inwardly, they look at others with friendlier eyes and now display a more lively interest in them."

Doctor Good did not follow up this observation, and it seems that he attributed the changed behavior of his patients not so much to a direct influence of procaine, but rather to a cessation of pain which gave the sick and depressed people a psychological lift.

Doctor H. Warren Crow, Chief of the Charterhouse

Clinic, designates procaine therapy as "the most valuable weapon in the treatment of the individual rheumatic patient." Edematic swellings in the joints are also reduced, and the treatment often leads to a lowering of the weight of the patient—perhaps due to a loss of water, of which too much has been concentrated in the peri-articular and subcutaneous tissues. Since most arthritis sufferers are overweight, this loss of weight constitutes an additional gain.

At the Karlsruhe Therapy Congress in 1957, Prof. Aslan and Cornel David reported the results of a study of the effect of procaine therapy on degenerative joint diseases, as seen in 100 old men or prematurely aged patients. Ninety of these patients were hospitalized for from 30 to 120 days, but remained under observation for another three to four years. Ten of the patients were under the Institute's care for over seven years. All patients received 5 cc injections of 2 per cent procaine, at a pH of 4.0, according to the regular schedule of treatment.

Of the 100 cases thus treated, 28 showed significant improvement, 60 some improvement, and only 12 were unchanged.

"Cases judged to have 'significantly improved', according to Prof. Aslan, "included those where functional capacity had been restored, where both static and dynamic pain (pain in a state of rest or motion) had disappeared, where both active and passive motion had been regained, where physiological and biochemical tests showed a return to normal or in the direction of normal, and a few cases with remarkable restoration of normal skeletal structure. Cases adjudged to have 'improved' included patients with improved active and passive motion, with reduction in the duration and intensity of pain and return in the direction of normal of some physiological and biochemical criteria. Patients evidencing no significant effect of treatment were labelled 'unchanged'."

Professor Aslan's successes with procaine in arthritic patients are not as striking as those described by other physicians. Almost 20 years ago, on the basis of 40,000 procaine infiltrations in a variety of rheumatic complaints, Dr. E. Fenz reported 75 per cent cures, improvement in 15 per cent, and no change in only 10 per cent. Professor Aslan believes that the variable results she has achieved—and which are in line with the findings of Spiess more than fifty years ago and Lériche during the twenties—are due in part "to differences in central nervous reactivity, to differences in the reactivity in the organism, and various other complex internal factors, among which endocrine responses play a prominent role." In spite of this she advocates the broad application of procaine therapy in cases of osteoarthritis.

The New York physician Dr. David J. Graubard,
already mentioned, has used intravenous procaine
therapy in rheumatic patients for the last fifteen years.
He and his co-workers administered intravenous procaine infusions plus Vitamin C either daily or weekly,
and reported considerable improvement in most cases
of rheumatoid, traumatic and osteoarthritis. In rheumatoid arthritis the pains and muscle spasms have
been subsiding to a point where physical therapy
could be begun. Best results were in osteoarthritis,
where not only was painlessness achieved, but mobility of the joints also considerably improved. Only very
advanced cases did not respond to this treatment.

### Osteoporosis

Another malady commonly associated with old age,
but occasionally occurring in younger people and even
in children, is osteoporosis—the decalcification of the
bones. It, too, has been combated successfully with
procaine therapy. From 20 to 30 per cent of all old
people suffer from this rather painful chronic disease,
which also makes them extremely prone to fractures,
particularly of the ribs and thigh. Sometimes younger
women are affected by osteoporosis after surgical or
X-ray castration, which indicates the connection between this illness and the functioning of the ovaries.

The standard treatment today is a high calcium diet and the administration of androgens and estrogens (male and female sex hormones).

At the Institute I saw two small girls suffering from this disease. In one, the case was diagnosed at the age of six, when it was found that she was unable to write. She underwent three years of unsuccessful treatment before coming to the Institute. After four years of procaine injections, she is now considered completely cured at the age of 13. Physically she is still underdeveloped, in relation to her age, but not to the degree that she had been in earlier years. Her menses have not yet started.

The other case had been operated upon and her hands placed in casts, which only worsened her condition and led to an osteoporosis of disuse. After several years of procaine injections, she is now on her way to recovery.

According to pharmacological research conducted in recent years (particularly by Prof. Eichholtz of Heidelberg) certain diseases that were formerly treated with calcium, such as bronchial asthma, nettle rash (urticaria) and various skin edemas, respond just as well, if not better, to procaine injections. This would seem to indicate that procaine encourages the body's retention of calcium taken in through the diet.

# Chapter 18

## *Diseases of the skin, and allergies*

Perhaps one of the most striking sights in the old age home at the Institute is the repigmentation of hair that has occurred. This process is in different stages: in some people the roots are growing in according to their natural color, whereas the ends of the hair are still white; others, particularly blond people, once again possess a complete head of naturally colored hair. In others, whose hair was white when coming to the Institute, the hair has taken on a greyish tinge. Without exception, the procaine-treated patients have a healthy growth of hair, with any bald spots receding rapidly.

The trophicity of the skin is restored to almost normal: brown blotches disappear, wrinkles smooth out, and senile keratosis, ichthyosis and erythrodermia (the pathological reddening of the skin) are successfully combated. Other skin diseases also can be treated with procaine; among them scleroderma, leucoderma (the appearance of white patches on the skin, often due to syphilis), vitiligo (a congenital pigmentation deficiency) and psoriasis, as well as simple rashes and eczemas. The growth of nails as well as paradentosis (degenerative inflammation of the gums) are influenced through procaine therapy.

### Alopecia

The therapeutic value of procaine in this respect is indicated by the case of Maria Tabarcea, who came to the clinic early in 1955 with a total alopecia (baldness). After a little more than two years of procaine injections her hair had completely regrown, she was discharged as cured, and according to the doctors who have since observed her, her hair growth remains normal. "Before and after" photographs of Maria appear in the photo section (pp. 81-96).

In Bucharest, I also saw an 18-year-old boy who some time ago lost not only the hair from his head, but from his body and face, including the eyebrows. After seven series of twelve injections, his body and facial hair had completely regrown. Since he still had two large bald patches on his head, the treatment in his case is being continued. In answer to my question, he said that he also felt better and stronger than before he lost his hair, and had better power of concentration and a greater capacity for learning (all of which, of course, could be merely a psychological reaction).

## Vitiligo

The success of procaine injections in vitiligo, sometimes called piebald skin to characterize the discoloration of the skin on the face (and sometimes the hair), is indeed impressive: under the influence of procaine, the white patches disappear. While the occasional spontaneous disappearance of vitiligo has been reported, the curing of this badly disfiguring ailment with the help of procaine in almost every instance would seem to be indicative of the involvement of a genuine healing process.

## Scleroderma

One of the worst diseases known (fortunately quite rare) is scleroderma, also called sclerosis of the skin. While it is classified as a skin disease, it results ultimately in paralysis of certain muscles, ulceration of bones, and finally, complete invalidism and death, since so far no effective treatment has been found. At the Bucharest Institute 22 cases of scleroderma have been treated, enough to warrant a report which will be published soon. I have seen one woman, now 37 years old, who has been under procaine treatment for three years, after having suffered from scleroderma for almost two decades. When she was brought to the Institute by her mother, the skin over her entire body had the color and consistency of wood, her face was almost expressionless, her nose was paper-thin, her finger bones (phalanxes) were ulcerated and about to fall off. Her knee joints were completely stiff, she was unable to open her mouth, was suffering from advanced paradentosis, had to endure excruciating pain and, according to the notation in her clinical record, was moribund. Since some of the internal organs were also affected, she had to be fed intravenously. For the most part she was completely apathetic. All possible forms of therapy had been used, including cortisone, but the disease continued to progress rapidly. She was brought to the Institute as a last resort.

Procaine treatments were begun immediately, at first according to the usual scheme, and then daily, but with an interval after every twelve injections. Today, she is no longer in the Institute's clinic, but is an outpatient. Her body skin has regained full trophicity, except for that on her hands and forearms, which are still thin and emaciated, and which she can move only a little. Being a seamstress, however, she has recently begun to sew again with slow, narrow motions. She can talk for the first time in years; the paralysis of the mouth region is gone so that she is able to feed herself.

She is without pain and walks, albeit with the help of a cane since her knee joints are still weak and will not support her for any length of time. She is still far from being a healthy person—but she has come a long way since 1956, and today feels optimistic. Since she came to the Institute with such a long-standing case of scleroderma, the doctors do not completely share her confidence; nevertheless, it is most remarkable to hear about the transformation this woman has undergone. Several other cases of scleroderma now under treatment in the clinic are of relatively recent origin and are given a better prognosis. In view of the initial successes with this disease, new cases in Rumania are now immediately referred to Prof. Aslan, a procedure which gives them a much better chance of recovery.

**Psoriasis**

The most dramatic evidence that procaine can be effective in curing psoriasis is the case of Tanasula Mircea, who had been suffering with psoriasis for 17 years, along with painful joint conditions which made him, at 42, a bedridden old man. He had been treated for five years with other drugs and with X-rays, to no avail. After 24 injections of procaine he was able to walk again, and the psoriasis was clearing up. The change is shown in the photo section (pp. 81-96).

Again, the procaine treatment of psoriasis is no invention of Prof. Aslan. The first successful instances of such cures were reported in the literature about twenty years ago. In the case of this dermatological condition, in which many parts of the body are covered with reddish, dry patches and greyish white scales, cure means a *complete and permanent* disappearance of the symptoms. Psoriasis has a tendency to improve spontaneously during the summer months, only to return with renewed vigor in the Fall, for as long as symptomatic treatment is the only therapy used. Many cases of what seem to be true cures are recorded in the Institute's files.

### Ichthyosis

Ichthyosis, the so-called fishskin disease, also has a good chance of being subdued by procaine injections, although it is generally considered incurable. This is an illness which manifests itself in earliest youth by keratinization of the skin, which usually lasts throughout life.

The Institute reports that in the case of a 6½-year-old girl suffering from hereditary muscular atony and ichthyosis (neither condition had responded either to other drugs or to physical methods of treatment) 60 injections of procaine produced not only a considerable alleviation of the muscular atony, but the reappearance of a trophic skin.

**Bronchial asthma**

Because bronchial asthma, like so many skin diseases, has been thought of primarily as an allergic reaction of the body, we will consider the effects of procaine therapy on this common affliction in this same chapter. As we have noted, Prof. Aslan was using procaine to relieve asthma attacks when she first became interested in the therapeutic possibilities of this drug. Some Soviet research, however, evidently predates the Bucharest work in this area.

Dr. N. K. Gorbadei reports that he began to use intra-arterial infusions of 0.5 per cent procaine solution with penicillin in the treatment of patients with bronchial asthma in 1943. This method has been tried in 32 patients (11 males and 21 females), 25 of whom had suffered from the complaint for over four years. In 17 patients attacks occurred over five times a day, and 11 had status asthmaticus (extremely severe attacks lasting from a few days to a week, sometimes fatal).

After four or five intra-arterial infusions of procaine and penicillin the general condition of the patients improved, the shortness of breath was relieved, the attacks of asthma were prevented, and the appetite and sleep improved in 30 of the 32 patients. In the remaining two patients, the attacks, although not completely abolished, became less frequent and less severe. Coughing was diminished and the normal rhythm of respiration was restored. Accompanying inflammatory changes in the respiratory organs were rapidly resolved.

Again, as Prof. Aslan emphasizes, she had made a 'rediscovery'.

# Chapter 19

## Diseases of the cardiovascular system

The long-term treatment of cardiac patients with procaine (usually for at least eighteen months) will often yield satisfactory results. Procaine has a vasodilatory effect, causing blood vessels to dilate, thus lowering the blood pressure. Arteriosclerotic conditions are improved, angina pectoris attacks diminish, heart thromboses show signs of faster healing. Intravenous injections have a favorable effect on cerebral spasms in older people; phlebitis, the recurrent inflammation of a vein, is easily overcome, and disturbances in peripheral circulation (after intra-arterial injections) are lessened.

The effects of procaine on these conditions have long been known and often described. In heart arrhythmies, treatment with procaine is standard procedure (there are in this country even oral procaine preparations available for this purpose). In other heart conditions procaine was prescribed as long ago as 1924.

The Soviet researcher N. K. Gorbadei, of the Leningrad Sanitary-Hygiene Medical Institute, examined over 70 of his peptic ulcer patients electrocardiographically before and after intra-arterial infusion of procaine. The general conclusions which Dr. Gorbadei derived from these examinations are:

Procaine, administered by intra-arterial infusion, has no toxic effect on the contraction of the heart muscle. The electrocardiographic studies of the patients seem to suggest that procaine infusions have a general action on the body as a whole, including the cardiovascular system, which is able to adjust temporary imbalances in the relative proportions of the excitation and inhibition of the nervous system.

The condition of the cardiovascular system before and after procaine infusions was studied by determining the tension of the veins and arteries by a bloodless method—measuring the venous and arterial pressures, and recording simultaneously the changes in the volume of the limb and the blood pressure in ink with a device called the kymograph.

The initial values of the arterial and venous pressures in the peptic ulcer patients were rather low, and the tension of the blood vessels was changeable, with a tendency to be raised.

After the first intra-arterial infusion of procaine a slight fall in the arterial and venous pressures was observed, presumably due to a fall in vascular tension as a result of procaine block.

However, after completion of the course of procaine infusions, most patients had higher arterial and venous pressures. The blood flow from the heart into the limb was increased, and the pulse rate slightly quickened. The limb volume was increased because of the greater influx of blood. The tension of the blood vessels became more stable. In other words, the intra-arterial infusions of procaine had a normalizing effect on the vascular tone and on temporarily deranged reactions of the cardiovascular system. Subsequent infusions of procaine had much less effect than the earlier ones.

In peptic ulceration, changes in the cardiovascular system are present, in addition to the gastrointestinal disturbances, and can be recorded plethysmographically. These changes respond to intra-arterial infusions of procaine along with the improvement in the course of the ulcer itself. The conclusion is that this is a rational form of treatment, which acts on the body via the receptors of the blood vessels and the central nervous system.

### Angina pectoris

At the same Institute, intra-arterial infusions of procaine were given to 20 patients (15 males and five females), aged from 20 to 71 years, who were suffering from angina pectoris—in most cases associated with early hypertension. (The patients were selected to exclude those with myocardial infarction.) All the patients were suffering frequent attacks of intense pain. The usual number of procaine infusions given was from four to ten.

After the treatment, 17 of the 20 patients were completely relieved of pain in the region of the heart, and the pain was less frequent in the other three. The frequency of the occurrence of palpitations and shortness of breath in the patients was also considerably reduced. The electrocardiographic findings showed improvement, the heart rate became normal, and extra systoles were abolished.

Similar results were reported in 1956 by F. F. Kil'matova of the Kazan State Institute for the Advanced Training of Physicians, who also used this method for relief of the pains of angina pectoris.

### Varicose veins

Varicose veins is another of the diseases commonly associated with old age against which Prof. Aslan has found procaine injections to be a valuable medication. Other doctors had used procaine for the same purpose as long as twenty-five years ago. The technique requires procaine injections around the affected vein or into the femoral artery. Here, too, procaine seems to have true curative power. Not only are the dull aches of the varicose veins relieved, but the condition itself, which is caused by a breakdown of the valves in the vein with subsequent unsightly enlargement, is favorably affected.

Prof. Aslan reports a number of cures of varicose veins without recurrence provided the causative circumstances (long periods of standing or sitting upright, or heavy lifting) were eliminated.

Why procaine should have such an effect on the veins is not clear. Some researchers believe there is some connection between the varicose veins that appear with increasing age and the increasing failure of the endocrine glands to function properly. In view of the indications that procaine influences favorably less active endocrine glands, we may have an explanation why it is effective in many cases of varicose veins.

It is also interesting to note that in several patients with elephantiasis-like swelling of the lower extremities, caused by circulatory disturbances, improvement was noted after intramuscular procaine injections. (Sometimes elephantiasis is caused by the presence of parasitic worms in the lymphatic system, in which case procaine will be of no help.)

# Chapter 20

*Diseases of the gastrointestinal system*

All patients treated with procaine have a markedly
better appetite which, of course, may be an indirect
result of their increased vitality. Procaine also seems
to have a normalizing influence on the intestinal flora,
but this has not yet been sufficiently investigated to
allow any positive claims.

**Ulcers**

The most successful use of procaine in this area has
been in the treatment of stomach and duodenal ulcers.
Usually six intravenous injections suffice to stop the
pain, and the ulcers themselves disappear after 24 injections, in most cases permanently. Only one of the
Institute's ulcer patients had to be operated upon, and
this proved to be for a calloused ulcer. Procaine therapy in ulcers was first recommended more than three
decades ago, mainly by French and Belgian physicians.
It is thought that the involuntary nervous system is

affected by procaine, which in turn influences the etiology of ulcers.

Professor Aslan's chief assistant, Dr. Cornel David,
reports that he himself had been suffering from gastric
ulcers, which became completely quiescent after only
five injections. In order to prevent recurrence, the Institute recommends that former ulcer patients undergo
a prophylactic series of injections each spring and fall.
Prof. Aslan's therapy is merely a rediscovery in this
case as well. It was reported in the literature as far
back as twenty-five years ago that up to two-thirds of
all gastric ulcers were cured by use of procaine, as evidenced by X-ray data.

The Soviet researcher N. K. Gorbadei, reporting on
procaine treatment of 171 patients with gastric ulcers,
tells of rapid relief from the pain of the ulcer, normalization of the secretory and motor activity of the gastrointestinal tract, and disappearance of the dyspepsia.
Objective evidence of the value of the treatment was
found by X-ray, electrocardiographic, and plethysmo-
graphic findings.

A typical case history cited by Dr. Gorbadei tells of
a female patient, aged 48 years, who was admitted to
the hospital in 1954. The diagnosis was an acute stage
of peptic ulceration, gastritis, and periduodenitis. She
had been ill since 1951. X-ray examination on admis-

sion showed a duodenal ulcer crater measuring 0.3 x 0.3 cm, which was tender on being touched. After the fifth intra-arterial infusion of procaine, the patient was completely free from pains in the stomach, flatulence, heartburn, nausea and vomiting; her appetite improved, and the constipation, which had previously afflicted her for five to six days at a time, was relieved. During this time she gained over four pounds in weight.

One month later, further fluoroscopy and radiography of this patient showed no ulcer.

On the basis of more than 5000 intra-arterial infusions of procaine into the femoral arteries, Dr. Gorbadei and his co-workers consider that this technique is safe, simple, and more effective in the treatment of ulcers than the administration of procaine by any other method.

# Chapter 21

## *Effects on the endocrine glands*

Many of the changes observed by Prof. Aslan in the procaine-treated old people indicate that this substance acts upon the endocrine glands. Hair growth, as already mentioned, is stimulated; some testicular function, often almost dormant, is revived; small amounts of estrogen (the female sex hormone) are found circulating in old women again (the return of pigmentation to the labia minora is only another sign of estrogenic stimulation); the adrenal glands become more active. The involution of female genitals usually is halted after several years' treatment, and in a few cases has even been reversed.

That procaine has a stimulating effect on the glands is also indicated by the fact that some doctors do not recommend use of procaine therapy in women prior to the menopause. This question has not been finally settled as yet, mainly because younger women have not been treated extensively, nor have any tests been conducted on experimental animals. Several cases of amenorrhea (absence of menstruation) in women about 40 years old have yielded after one or two series of procaine injections, with regular menstrual periods appearing thereafter. Several cases of correction of sterility, which had persisted for many years in women with normal organs, are also on record. Procaine, without the addition of any hormones, was found to be of help.

Laboratory investigations are now under way at the Bucharest Institute to verify the clinical findings that procaine stimulates the production of adrenocortical hormones. Tests, first on mice, then on men, have already shown that procaine inhibits the thyroid function, restoring this gland to normalcy in cases of overactivity. Much still has to be learned about the total effects of procaine on the endocrine glands. It will be interesting, indeed, if tests prove that procaine as well as aspirin, other analgesics, the cortical hormones, rutin (Vitamin P), calcium salt and antihistaminic substances, has the property of warding off the brittleness of capillary vessels.

In this respect, some wartime research conducted by Dr. Georges Ungar at Oxford University is of considerable interest. Dr. Ungar found many different procedures could protect experimental animals against traumatic shock. These procedures included the production of previous minor traumas, and the administration of ascorbic acid, procaine, cocaine, adrenaline, a whole cortical extract, an extract of the pituitary gland, and especially—the adrenocorticotrophic hormone (ACTH).

This researcher, who is now Director of the Department of Pharmacology at U. S. Vitamin Corporation, found also that the blood of healthy persons contains a substance which he believes is produced by the spleen, and which he called 'Splenin A'. The blood of persons suffering from rheumatoid arthritis contains a quite different substance, which he called 'Splenin B'. Patients who have recovered from the disease again have 'Splenin A' in their blood.

Dr. Ungar found that under the influence of ACTH (the anti-stress hormone from the pituitary gland) the amount of 'Splenin A' in the blood of experimental animals increases, while 'Splenin B', if it is present, decreases. By ingenious experiments on guinea pigs and rats, he was also able to show that procaine, like 'Splenin A', reduces the bleeding time and increases the strength and resistance of the capillaries. Verification of his research on a larger scale in 1952, reported by the British endocrinologist Dr. Raymond Green, indicated that procaine, like cortisone and ACTH, is capable of calling out the anti-stress hormones from the pituitary glands and the adrenal cortex, and also shows us how long ago and how widespread was British interest in the potentialities of procaine in the treatment of old age diseases.

# Chapter 22

## *Why procaine was "forgotten"*

There is a well-known quip to squelch the braggart who is too enthusiastic about his own intelligence: "If you're so smart—why ain't you rich?"

In effect, we must ask the same question about procaine. If it is such an effective remedy in so many diseases, if it has been known to be effective in these diseases for as long as fifty years, why isn't it being used by every doctor today? Why does the average practitioner still think of procaine only as a local anesthetic? To answer these questions, we need to look back at the medical headlines of ten years ago.

In a story entitled "Why Britain was wrong to drop it," the *'Daily Mail Doctor'* writes in that London paper as follows:

> "When British doctors first began to show interest in the novocain treatment of asthma and rheumatism (just after the second world war ended) shattering news came out of the Mayo clinic in America. Cortisone had been isolated as a drug, and was being used to make cripples walk.

"Dr. Hench, the man responsible, came to London in 1950 and was received like an Eastern potentate at the Royal Society of Medicine. The medical press hailed him. . . . So procaine was dropped as being of little interest now that the wonderous cortisone was available. . . .

"But ten years later the picture looks very different. The rosy promise of cortisone itself and of ACTH (the pituitary hormone which stimulates the flow of cortisone from the adrenals) has not been fulfilled. It has turned out to be a two-edged weapon—curing some symptoms in the stress disorders, but making others very much worse. . . ."

The *'Daily Mail Doctor'* writes as if his personal experience with cortisone had been bitter—and bitter indeed was the disappointment of the world of medical science as it became apparent that the side effects and the long-delayed after-effects of cortisone in so many cases outweighed the benefits of this drug.

Hindsight is easy. The knowledge that has come with experience in use of cortisone in no way detracts from the accomplishment of its isolation. In its preparation and use, biochemists and medical researchers have learned much that will be of great value in the future of medicine and chemotherapy.

ACTH and cortisone are able to control such often excruciatingly painful states as rheumatic fever, arthritis, asthma, Addison's disease, certain allergies, etc. But if the treatment is discontinued, the symptoms of these diseases usually return. Moreover, the side effects often prove very disturbing: personality changes, gastric difficulties, and the syndrome of Cushing's disease (accumulation of fat in the face, the abdomen, and the buttocks) have been observed.

The medical profession is now fully alerted to the danger of overenthusiastic and undercautious use of cortisone and ACTH, and these drugs are used today only under the most careful supervision, so that patients will not have to suffer the untoward side effects.

Possibly the most unfortunate side effect of cortisone was the way in which excitement about its discovery cut off the growing medical interest in procaine as more than a local anesthetic.

Cortisone, as well as ACTH, has an antiphlogistic effect—which as long ago as 1906 Prof. Spiess claimed for procaine. The first two drugs do their work much more rapidly and dramatically, and indeed, are often indispensable. Procaine is less spectacular and requires a longer use before there are any signs of success, but it has virtually no side effects.

And just as medical science does not yet know how cortisone, ACTH, or even the time-proven aspirin work in the human body, so it has not yet solved the mystery of procaine's action.

# Chapter 23

## *"H₃"—a name for a riddle*

The many changes that procaine brings about in the
sick and aged body lead, of course, to the question of
how this substance acts. Is procaine merely a catalyst?
Does it combine with other substances to form some
new compound? It has been suggested that procaine
does not enter into the reaction at all, but merely frees
certain enzymes. In spite of the efforts of many pharmacologists and biochemists, no single definitive answer has been evolved. Nor do we know why it should
come closer to revitalizing old people than any other
drug known, or why it should act as a psychic ener-
gizer in young people as well.

The current use of procaine, although limited, is
actually a reversal of established laboratory techniques, wherein the substance under investigation
must be thoroughly analyzed before it is put into use.
However, strictly empirical therapy is not a new thing.
Another important therapeutic agent, aspirin, has not

been explored sufficiently to uncover its mechanism and mode of action.

The parallel between procaine and aspirin is striking indeed. Both of these familiar drugs relieve inflammation and pain, and both reduce the crippling effects of arthritis. Aspirin, like procaine, improves the circulation of the blood. Furthermore, neither of the two drugs is habit forming.

Procaine parallels aspirin in another way that may be of considerable importance if procaine proves to be as effective against old age as Prof. Aslan claims—it is not an expensive substance to produce!

While, as we have mentioned, procaine has been of value in over 150 serious diseases and lesser ailments (not to speak of its importance in geriatric practice), aspirin has been used (mostly for symptomatic treatment) against some 200 illnesses. The mystery that surrounds the impressive histories of both these substances is deepened by the fact that medical science does not know why either procaine or aspirin act the way they do.

Procaine, we may briefly repeat, is hydrolyzed in the body into two substances, para-aminobenzoic acid and diethylamino-ethanol. Both have been extensively studied but in themselves do not have all the effects procaine exerts. Para-aminobenzoic acid (PABA) was found to be necessary for normal pigmentation of hair in the rodent, maintenance of a normal fur coat in the rat, and for the multiplication of certain strains of bacteria. It is present in yeast and liver extract, and is considered a member of the Vitamin B complex. As such it has been given the name $H_1$. But so far it has been impossible to determine PABA's exact role in the human body; we do not know whether it is necessary either for the body's nourishment or its functioning.

Diethylamino-ethanol (DAE), the other component of procaine, is believed to be more responsible for the properties of the parent drug than PABA, but in itself does not account for the beneficial effects of procaine. DAE, according to pharmacological textbooks, induces local anesthesia, exerts a quinidine-like action on the heart, a spasmolytic effect on smooth muscle, and also posseses analgesic and antiallergic action. In particular it has proven its therapeutic value in the treatment of heart arrhythmies, where it is thought to be even more effective than procaine, although it must be administered intravenously in considerably higher doses. This is of no concern, however, since DAE is even less toxic than the almost nontoxic procaine.

Neither PABA nor DAE has a trophic effect on the skin, nor do they stimulate the general metabolism of the body (leading to an increase in weight) as strongly as does procaine. For that reason it must be assumed that the procaine molecule as a whole must have a very specific effect. Whether this is vitamin-like, as Prof. Aslan believes, whether it has something to do with the production of acetylcholine in the cell structure, as other researchers surmise, or whether a direct influence on the central nervous system is involved, has not yet been proved. Perhaps it is a combination of two or more of these actions.

G. N. Udintsev and V. B. Blank of the Leningrad Sanitary-Hygiene Medical Institute reported in 1957 on their study of the morphological changes in the blood of 150 peptic ulcer patients, before and after procaine infusions. Their results revealed that there is no toxic action of procaine on the bone marrow, and they believe that the changes in the blood produced by procaine are brought about by reflex redistribution of the blood in the body and by reflex action on the bone marrow and hemopoietic system generally.

Soviet researchers have been actively studying and reporting on the effect of procaine on the body, and the mechanism of its action for as long as 20 years. The recent book by N. K. Gorbadei discusses applications of procaine in the treatment of peptic ulcers, and reviews the theories of the mechanism of action of procaine. After citing the theories put forth by various

Soviet biochemists and physiologists as long ago as 1934, Gorbadei continues:

"On the basis of the foregoing we may conclude that no unanimity yet exists with regard to the interpretation of the mechanism of action of procaine. In the opinion of A. V. Vishnevskii, A.D. Speranskii, and N. I. Leporskii, by whatever means it is administered, in addition to its local action procaine also has a general action on the body. This general action of the drug is due to normalization of the processes of inhibition and excitation in the central nervous system.

"Our experimental and clinical findings give grounds for the assertion that when given by intra-arterial infusion, besides its action on the vascular receptor apparatus, procaine also has a general action on the body as a whole, through the central nervous system. The intra-arterial infusion of procaine thus has an indirect action on the body via the neurovascular receptor apparatus and the central nervous system. The action of procaine undoubtedly depends on the initial functional state of the nervous system, and on the mode of administration. The diversity of the functional changes resulting from procaine administration, the speed of the reactions to it, and the manifestation of the effect far from the site of injection all underline the role of the general, evidently reflex, action of procaine rather than its local action. This is in agreement with the findings of Andreeva, Komarov, and Timeskov (1957), who consider that the therapeutic effect of visceral anesthesia is mainly dependent on the general neurotrophic action of procaine on the body.

"Most authors thus consider that procaine acts through the nervous system. At this point, however, their views diverge . . .

"From the data in the literature and our clinical observations, there are grounds for believing that an essential role in the mechanism of the intra-arterial infusions of procaine is played by nervous reflex influences arising from the receptor apparatus of the blood vessels, as well as by humoral factors, after the entry of the procaine into the blood stream."

As was to be expected, Prof. Aslan herself, wanting to know more about the mechanism of procaine, has done extensive research at her Institute, particularly on plants and animals. This is not the place to describe the highly technical experiments which, while interesting in themselves, have not led to more than a hypothesis as to why procaine brings about the mentioned effects in human beings. Her conclusion: that procaine has a distinct vitaminlike effect, but that it may also act as a biocatalyst (speeding up or slowing down certain chemical processes). It was in order to distinguish it from PAB that she proposed to call procaine "H₃," which is the official name of the buffered procaine she is using in the treatment of her patients.

That the vitaminlike effect of procaine is the answer to the riddle may be doubted in view of the fact that, as a rule, procaine is much more successful in the treatment of old age than even extended vitamin therapy. Also the signs of "rejuvenation" appear much faster than in people largely treated with vitamins. An explanation, if one is possible, should also be forthcoming as to why the few milligrams of PABA set free when the injected procaine hydrolyzes should be able to maintain in a patient a more youthful appearance for months, as sometimes happens after only one or a few injections of procaine.

As an aside, it is interesting to note that the coca leaves commonly chewed by Peruvians and Bolivians living on high mountains have some of the effects now ascribed to procaine. Cocaine has been derived from the coca leaves and, as we have learned, procaine was synthesized as a substitute for the toxic cocaine, which is habit forming. Strangely enough, the coca leaves do not have this effect; when coca chewers move to the cities, they usually stop using it. While digesting coca leaves, they drink and smoke little (as a matter of fact, infusions of coca capsules are being used for breaking the drinking and smoking habits without creating the danger of addiction). According to medical investigations, coca acts as a heart tonic, stimulates the contractions of the heart, thereby enriching the body with oxygen, and increases the excretion of nitrogen, chlorides, sulfates and phosphates in the urine, which may be the main reasons for the sense of well-being engendered when it is chewed. Its anesthetic effects on the palate and the mouth alleviate feelings of hunger and thirst, enabling coca chewers to walk up to fifty miles daily without fatigue and without any intake of food. Tests made on Indians during and immediately after their coca chewing showed an increase in the metabolic rate. This rate remains high with chronic coca users, clearly showing a stimulation of the metabolism as a whole. No observations have been made as to whether coca may also have some eutrophic effects.

# Chapter 24

## *DMAE—a related mystery*

While little original pharmacological work with procaine has been done in this country, some very interesting information has come to light during the past two years concerning the action of dimethylamino-ethanol (DMAE), a slightly changed form of diethyl-amino-ethanol (DAE). Prof. Carl Pfeiffer, professor of pharmacology at Emory University in Atlanta, Ga., and his co-workers have reported their findings in *Science* (July, 1957) and the *Journal of Pharmacology and Experimental Therapeutics* (1958). Since their research may contain a clue which will help researchers answer the riddle of how procaine works, we summarize it here.

Daily oral doses of 10 to 20 mg DMAE within seven to ten days produce a mild and pleasant degree of central nervous stimulation, which is characterized by less fatigue and sounder sleep. Also, fewer hours of sleep are needed. Larger doses may result in increased muscle tone but may also produce insomnia. The stimulation of the central nervous system is not accompanied by a rise in blood pressure, a rise in body temperature, or a change in the plasma level of protein-bound iodine.

The similarity between DMAE and the DAE component of procaine and the similarity of the effects produced by these two compounds in the human body would indicate that when medical science learns how one works, it will also understand the mechanism of action of the other.

In Prof. Pfeiffer's second paper he discussed a double-blind study, comparing DMAE therapy to a placebo. A questionnaire was used to supplement weekly measurements of heart rate, blood pressure, muscle strength, hand steadiness, vital capacity and body weight. This therapy continued for three months, and during the last six weeks all students were being treated with DMAE. In Prof. Pfeiffer's own words:

> "Significant subjective changes found in the DMAE-treated group were increased muscle tone, increased mental concentration, changes in sleep habits. In most instances the sleep habit was less sleep required. Others

reported sounder sleep with earlier, clear-minded awakening. A mood change to greater affability or mild euphoria was coupled with a more outgoing or outspoken personality. No significant changes occurred in heart rate, blood pressure, muscle strength, hand·steadiness, vital capacity and body weight.... Twenty-five out of the 35 students noted mental stimulation, which increased daily in the first week of medication and was greater than that produced by amphetamine [a so-called "pep" pill]. Five students discerned no effect at the dosage used. Unlike that produced by amphetamine, the DMAE stimulation lasted 24 to 48 hours after discontinuation of the dosage, and was not accompanied by a rebound period of depression. An overdosage produced insomnia, muscle tenseness and spontaneous isolated muscle twitches."

Other researchers have since reported similar effects with a DMAE drug called Deanol. Two members of the University of Washington School of Medicine, in Seattle, have stated that in a group of 100 patients suffering from various psychiatric disorders, especially exhaustion and depression, increased energy and lessened depression were noted in most cases after the initiation of the therapy with Deanol. Improvement usually occurs within a few days and no side effects are observed, except for occasional overstimulation, which is controlled by a reduction in dosage.

Two years of research after the initial observation have not led to an explanation for the mode of action of DMAE. Prof. Pfeiffer theorizes that it acted as a precursor to acetylcholine, increasing the amount of the latter in the brain, from where it would be transmitted to the nerves. (Acetylcholine is constantly produced and destroyed in the parasympathetic nervous system, and the latter's tonus depends upon an effective concentration of acetylcholine in the cell.)

# Chapter 25

## *Can procaine postpone old age?*

The most dramatic aspect of the procaine therapy is the possibility of its use as a prophylactic measure, to ward off the symptoms of old age, or at least to postpone them. On the surface it would seem that this could be the case, but experimentally no proof has been established. The complete lack of old age criteria will also make it difficult to prove or disprove the prophylactic effectiveness of procaine in human beings.

We have no way of determining the physiologically "normal" state of a 65-year-old man: we do not know how much sickness to consider "normal" in a man of that age. At best we may arrive at certain statistical data by examining a large number of 65-year-old people and striking an average, but biologically we are still in the dark as to the true picture each man or woman should present at a certain age. With this in mind, we can readily see how difficult it will be to gauge any prophylactic effect of procaine in an objective manner. Suppose we start a group of 45-year-old people on a modified procaine therapy, as has been done in Bucharest and elsewhere, and treat another group of the same age with another substance or with a placebo. Can we really set up standards of what to expect in these people, provided even that their physiological conditions at the beginning of this experiment are approximately the same? The subjective story might be different—an absence of vague complaints, a less frequent occurrence of nervous and joint diseases in the procaine-treated group would make the value of such prophylactic treatment clear.

The prophylactic use of procaine is not completely in the realm of speculation, as may be inferred from the fact that premature aging has been successfully reversed. We have heard of the particularly striking case of the 42-year-old man; similar but not quite so dramatic cases are to be seen almost daily at the Institute of Geriatrics. One Bucharest patient was observed in whom further aging seemed to have been arrested for a period of seven years.

We have mentioned that the Geriatrics Institute has under its care a surprisingly large number of children with either unusual diseases ( osteoporosis, ichthyosis, etc.) or with more common ailments, such as asthma and rheumatism, which respond readily to procaine.

Cases of the latter type are explained by Prof. Aslan as instances of precocious aging of the organism. She feels that neither bronchial asthma nor rheumatism will normally occur in children, and that we are dealing in such cases with relatively rare symptoms of old age in the young. (In girls, procaine therapy must be administered very judiciously, since procaine stimulates the ovaries. The potential benefits must outweigh the possible danger of hormonal overstimulation.)

While procaine has proved itself in the simpler cases, it is recommended that the treated children continue periodical procaine therapy in order to prevent any recurrence of these untimely signs of old age. The state of health in these children in the later years will be of great interest to medical science.

It should also be interesting to observe whether procaine can be of any value in a very rare condition called "progeria," a form of senility which is apparent by the end of the first year of life. Very few cases of progeria have been described, and all of them have led to death before the 26th year of the affected individuals. These children show atrophic skin, baldness, calcification of the vessels, osteoarthritis, osteoporosis, sexual retardation, tremors of the limbs — in other words, all the signs of senescence with the exception of cataracts. Death is usually due to coronary insufficiency. The cause of progeria is still unknown, but since procaine favorably affects most of its symptoms, it would seem logical to institute this therapy in cases of such accelerated aging.

The researchers in Bucharest are also naturally eager to know whether or not procaine can postpone or avert the symptoms of old age. Because of the harmlessness of this drug, the Institute is giving prophylactic treatments to several thousand people over 45. Animal experiments begun late in 1957 will be completed early in 1960. Prof. Aslan will be disappointed, but not surprised, if all these experiments indicate that procaine is not effective in postponing the onset of old age signs. It may well be as Soviet researchers have suggested, that procaine affects only an out-of-balance nervous system, and thus has little or no preventive effect. The caution which Prof. Aslan has expressed in this respect is in sharp contrast to her conviction, born of experience, that procaine can halt and reverse premature aging.

# Chapter 26

## *The Western doctor: "I'm from Missouri!"*

The medical profession in the West does not share Prof. Aslan's conviction that procaine can halt premature aging. It is perhaps even less convinced that a general therapy for old age has been developed in Bucharest. In part, this skepticism is undoubtedly a reaction to Prof. Aslan's unfortunate use of the word "rejuvenation." Also, many doctors in the United States are still under the impression that Prof. Aslan reported her work to the press rather than to her colleagues, whereas quite the contrary is true—she has always presented her reports to reputable medical bodies, including the IVth International Gerontological Congress in Merano, Italy in 1957.

There are, however, three very basic criticisms which medical researchers here make of the data on which Prof. Aslan bases her claims for procaine therapy.

**1.** The number of patients originally treated is too small for statistical analyses to be significant, and no careful statistical analysis of the thousands treated in the past two years has yet appeared. The medical statistician will turn to Table 1 on page ...., and point out, for example, that the deaths of only two more patients in any year would have increased the mortality rate by from 66.6 to 200 per cent! He will also note that the median age of patients who died is of no significance, especially since the median age of the patients able to work or able to take care of themselves is not given. He will want to know on what basis patients are admitted to the old age home; whether those whom doctors judge permanently helpless are admitted as readily as those for whom a reasonable prognosis is a continued period of self-sufficiency.

2. There is no indication that control groups were evenly matched. The table on page 69 indicates that the mortality of procaine-treated patients was only 2.7 per cent, and that of patients not treated with procaine was 10.3 per cent. But the careful medical scientist asks immediately: Were the patients in the procaine-treated group as ill as those who were not treated with procaine? Did this latter group of people receive any treatment at all? Were they receiving the same diet, given the same care, under the same conditions of sanitation?

Dr. Donald Mainland, in his book "Elementary Medical Statistics," tells of two researchers who were testing an extract of progesterone by injecting it into female rabbits. Their results were so successful that they were writing a paper about their work, when they decided to see if the less skillful chemist of the two could follow the directions they had written. He prepared a batch of extract and injected it into a number of rabbits. It did not work. He tried again, and the extracts again produced no effects. Finally the researchers made a batch of extract together, and each researcher injected it into a group of rabbits. In one group, the extract worked beautifully, in the second group—not at all.

At last they tracked down the explanation. Only a rabbit over eight weeks old would react to progesterone. The first chemist evidently thought heavy rabbits made nice experimental animals, and had invariably chosen animals old enough; the second chemist had an obvious subconscious preference for smaller rabbits, for none that he had chosen to inject was old enough to react.

If subjective factors can operate even in the choice of experimental animals, they can also affect a doctor's choice of a patient for experimentation.

Did Prof. Aslan and her associates, without realizing it, inject procaine into patients with more will to live, or with basically stronger constitutions? Only tests with carefully matched control groups can answer finally, one way or the other.

**3.** The importance of psychotherapeutic factors has been largely ignored in Bucharest. It is well known in geriatric medicine that an injection of even a simple saline solution or distilled water will often have a profoundly stimulating effect upon an aged patient. Furthermore, the enthusiasm and hopes of the doctor giving the injections also exert their effects, and the doctor's analysis of the patient's true condition is sometimes affected by what he hopes or expects to observe.

To rule out all of these psychic factors, medical science has devised the double-blind test: In this trial, a pharmaceutical laboratory supplies an equal number of doses of the medicine to be tested and of a placebo so identical in appearance that even the doctor who administers them cannot tell the difference. Subjective factors cannot operate in a trial of this sort, for neither the doctor, the patient, nor the person examining the patient knows whether or not a medicament has been given. Only after completion of the test or series of tests, when all the data has been collected, is the doctor informed which patient received the drug, which the placebo. And in many cases, the absolutely inert placebo has been as effective (sometimes even more so!) than the drug under test.

Even for researchers schooled in the most rigid modern methods, it is difficult to design a method of testing procaine injections as an old age therapy which will provide conclusive proof or disproof. Dr. Stanley R. Mohler of the Center for Aging Research of the National Institutes of Health writes that

> ... A few applications for research grants to study the effects of procaine on the process of aging have been received in the past by the National Institutes of Health. As of the present (October, 1959), no NIH grants have been awarded to these applicants because the advisory groups were not convinced that the experimental designs of the proposals were sufficiently planned and described. Certainly, NIH continues to be interested in supporting any proposal which relates to problems of health, including proposals to evaluate what is referred to by some as "procaine therapy."

Thus far, Prof. Aslan's claims are backed mainly by testimonial case histories, and, as *Lancet* reminds us, these have often misled medical investigators. Until more convincing evidence is brought forth, either by Prof. Aslan or by other medical workers, the Anglo-American attitude will undoubtedly be for extremely cautious, carefully controlled clinical trials, not for widespread application of procaine as a therapy for old age.

# Chapter 27

## Why all the excitement?

WE HAVE SEEN that there has been no miracle drug discovered at Bucharest, that "H₃" is merely a new name for something not fully understood about a very old and familiar drug — procaine hydrochloride. We have seen further that other doctors used procaine before (sometimes long before) Prof. Aslan, to treat exactly the same stress diseases. We have learned that there is still no sure answer to the riddle of how (or to what extent) procaine works in the human body, and we see that thus far there is no positive indication that procaine can postpone the normal onset of old age.

*Then why all the excitement?*

Because while others have applied procaine in the relief of one or another stress diseases (and stopped if the treatment was not promptly successful or as soon as some symptomatic relief was obtained), Prof. Aslan saw old age as a complex of these same stress diseases, and procaine as a potential weapon in the battle against old age itself.

Her reports that procaine is effective not merely as

a therapy against one or another disease, but against the debility, the helplessness, the childishness and the senility of old age, are tremendously exciting to the entire world, for old age strikes every man and woman who survives maturity.

In the United States alone, there are today more than fifteen million persons above the age of sixty-five. By 1980, this group may number close to twenty-five million. These figures point up the scope of a great national problem, for while aging may still be considered a crisis for the individual, it is today the concern of the communty as a whole.

That Washington today is aware of its responsibility in this respect is obvious. Congress has issued a call for a "White House Conference on Aging", to be held in January, 1961. The Congressional act setting up this conference reads:

> "In order to prevent the additional years of life given to us by our scientific development and abundant economy from becoming a prolonged period of dying, we must step up research on the physical, physchological and sociological factors in aging and in diseases common among middle-aged and older persons!"

So far, our main emphasis has been on improvement of the care for the elderly. Up to now, medical science has failed to come to grips with the task of treating old age. Nobody is more intimately aware of this situation than are those physicians who deal mostly with old people.

Says Dr. Edward J. Stieglitz of Washington, D.C.: "Health is a lot more than the absence of disease. Pediatrics has been making healthy children healthier. Geriatrics could do the same. The trouble is that doctors think entirely in terms of disease, and are ignoring their opportunities for making aging people healthier."

This is the opportunity which was not ignored by Prof. Aslan—the opportunity to search for a therapy which would make aging people healthier.

It has been remarked that with old age the good characteristics of man diminish, while the bad ones increase. It is this very manifestation which tends to make old people valueless, superfluous, expendable. By the same token, it makes them a burden to the younger generation, individually as well as collectively. "Most of us are likely to feel almost instinctively that youth is everything and that old age is a disaster", Dr. Michael M. Dacso, director of physical medicine and rehabilitation at New York's Goldwater Memorial Hospital, recently stated. "We pity old people. Often we avoid them. Sometimes we laugh at them." To our current way of thinking, as psychiatrist Dr. Jack Weinberg of Chicago's Michael Reese Hospital has so aptly put it, the aged are "simply people without a future."

Incidentally, while most medical schools in this country have a chair for pediatrics, none has as yet one for geriatrics. In spite of the strides made in recent years, the aged have remained medicine's stepchildren.

The attitude of defeatism which has been characteristic of so much of the world's effort on behalf of the elderly may be changed through the application of the procaine therapy *if it should prove as effective as Prof. Aslan and her collaborators believe.* For this is the first proposal for a treatment for old age which could be applied on a broad scale because it is simple to administer, inexpensive, not dangerous, and not habit-forming.

Western scientists still fear that Prof. Aslan is over enthusiastic about the results of her therapy. The British Medical Journal referred to her as a "woman gifted with humour, charm, enthusiasm, and (here was the rub) boundless therapeutic optimism." Yet even if only a small fraction of her enthusiasm for the procaine therapy proves justified, a tremendously exciting fact remains: *the battle against old age has been joined!*

## From the Editor
### — reveille

Possibly what we have just read is the story of a breakthrough the age barrier. More probably, the procaine therapy will prove to be one more valuable weapon in medicine's arsenal against disease. Time and rigid testing may even show it to be only one more medical dud.

The facts should emerge in 1960. Laboratory research, controlled tests, and clinical studies of this therapy are under way at the present time in both the United States and Great Britain. The results of such studies will be published in appropriate medical journals and reported to medical meetings, and will thus be available to all doctors.

Whatever the outcome, the exciting fact is that the battle against old age has begun, and the real question surely is whether we are going to have an all-out effort, or another Hundred Years' War.

If the present tempo and scope of Western research into old age and its diseases are not quickened and expanded, the prognosis is for a long drawn-out struggle. The delayed reaction of American (and British) medicine to the presentation of the claims for procaine therapy is a good case in point.

At this writing (January, 1960) two and one-half years have passed since Prof. Aslan reported her clinical results to an international medical congress. It has been fourteen months since Henry Marx first informed the American public of this therapy in a national magazine. Yet the definitive answer to the question: Does this therapy work? has still not been provided either by government research centers, medical societies, or by the research or medical departments of the pharmaceutical companies. In fact, there are not even tentative answers. The card catalog of the National Library of Medicine—one of the largest and most efficiently organized medical libraries in the world—does not list a single report of Western Research either in verification or in denial of the efficacy of this treatment, as regards the therapy of diseases of old age in humans, or as a prophylactic measure on the basis of animal experiments.

This is in spite of the fact that the diseases of old age against which Prof. Aslan reports procaine to be an effective treatment have reached epidemic proportions in the United States. Over 15 million Americans and their families — together surely at least a quarter of the population—are directly concerned.

It would be easy to lay the blame for the slow-motion approach to investigating the potential of procaine therapy on one or more of the usual scapegoats —government, the medical societies, or the practicing physician. But is not the real culprit a general attitude of fatalistic, apathetic acceptance of the ills, sufferings, debility and death of those over 65—an attitude that is prevalent even among the victims themselves? The man or woman who survives ninety years and remains a self-sufficient human being is regarded as somewhat of a curiosity, rather than as a living example of the potential of the human body. And the self-sufficient centenarian is considered a freak, not as having achieved naturally that which the medical profession must help us all to reach!

Suppose we considered that our medical science should achieve for humanity that which life has accomplished in a comparatively few cases unaided. Then surely no therapy which holds any hopes of improving either the quality or quantity of the years past 65 could be long untested. Research into death control might then take its place alongside research into birth control. And for each dollar spent with the aim of sending a man out of this world, we might spend a dime or so to keep ourselves in this world.

F. C.

————

*Do not go gentle into that good night.*
*Rage, rage against the dying of the light.*
—DYLAN THOMAS.

163

# Appendix

Case histories

Bibliography

Glossary of medical terms

# Case histories

*The first set of cases was reported in 1956.*

## Case No. 1
V. V.—Old age home patient, 91-year-old woman, 1.49 m tall, weight 42 kg, under observation for the past seven years.

*Diagnosis:*
Mental disturbances, hard of hearing, ichthyosis, senile keratosis, arteriosclerosis, hypertension, cutaneous epithelioma on the nose, extrapyramidal disturbances, acute arthritis of the right shoulder.

*Condition in May, 1949:*
Frequent hysterical crying, enuresis, cannot take care of herself, very weak memory (cannot remember the names of her dead children), muscular hypertonicity, bent body, takes only very small steps, completely white hair, clouded gaze, enophthalmus, many wrinkles in the face, particularly around the eyes, senility spots on face and hands, parchment-like facial skin, bleeding erosions of 1 cm diameter on the nose, on the chest and arms ichthyotic skin, general pruritus, trembling of hands, hearing and vision seriously impaired, weakness and slow patellar reflex, pronounced muscular atrophy, thinning of adipose layers, heart apex in the 6th intercostal space, pulse 104,

occasional extrasystoles, loud systolic murmur in the second right intercostal space, venosclerosis, blood pressure 210/140.

*Progress:*

She was treated for her acute arthritis with 10 cc of a 1 per cent novocain solution injected under the right shoulder blade at intervals of three days. After four injections the acute pains disappeared and her general state improved. On May 1, 1951, her condition had returned, except for the articular symptoms. The usual intramuscular treatment was then started. By July, 1951 her pruritus had disappeared and her general state had improved. In September, 1951, a receding of the extrapyramidal rigidity was observed. She became euphoric, her hearing and the condition of her skin had improved. The ichthyosis had disappeared almost completely.

*By December, 1951* her muscles had become stronger, her gait was improved. She held her body erect and was able to bend at the waist and touch the floor with her fingertips. She was able to take care of herself and to remember earlier experiences and dates. In April, 1952, she could take walks into the city by herself. She was also capable of such delicate actions as buttoning her clothes. By July, 1952, the reflexes came faster and more readily. By August 10, 1952, repigmentation of the hair could be seen on the temples. Trophicity of the nails had improved. the ichthyosis had disappeared completely. She was able to concentrate much better, her thinking had become clearer, its content more adult. Her general condition had become good. There was interest in caring for herself. The venosclerosis had disappeared, as had the rolling appearance of her veins.

*By October, 1953,* conditioned reflexes had become fixed after only three associations. The improvement in the general state of health was continuing. The wrinkles were smoothing out and her face had once again acquired color.

*Present condition:*

On *August 22, 1956,* the hair growth around the temples had become stronger and 80 per cent of its former color had returned. 20 to 30 per cent of the rest of her hair was repigmented—her eyelashes completely pigmented, her gaze vivacious. The occasional episodes of depression are gone; the old woman exhibits general liveliness. Several times a day she goes up and down the stairs of the Institute. Her muscles show considerable development. The skin on her legs feels smooth. The woman goes out by herself, remembers new experiences and recalls episodes of the distant past. Her weight has risen to 46.5 kg, and she is now 97 years old. Since the start of the treatment no new joint symptoms have appeared. Treatment with digitalis, at first needed twice a month, was discontinued in 1952, except when she was suffering from grippe or pneumonia (during the past five years this has occurred three times, twice during the winter). The cutaneous epithelioma shows deeply cicatrized layers. (This patient died in 1958.)

## Case No. 2

E.H.—Old age home patient, 70-year-old woman, treated since May, 1951.

*Diagnosis:*

Glaucoma and cataract, aortitis, arthrosis of the left knee, arteriosclerosis.

*Status ante:*

Since 1946 glaucoma in both eyes, the loss of vision leading to fits of depression.

*Status praesens:*

No more than three hours of sleep. Arthrosis of the left knee, slight extrapyramidal rigidity, pronounced amnesia, glaucoma and cataract on both eyes, atrophied skin, senile keratosis, gray temples, systolic murmur over aorta, brachial and carotid arteriosclerosis, blood pressure 155/65, pulse 54.

*Progress:*

By *January 8, 1952*, apoplexy with right hemiplegia and aphasia. Sphincter incontinence, subcomatose condition, one week persistent fever. During the coma, 0.10 g of procaine was injected intravenously twice a day. In the course of the injection the patient came to for about five to ten minutes. After the first ten days following the apoplexy, the conditions incident to hemiplegia began to disappear. The treatment was continued intramuscularly.

By *June 1, 1952*, the old woman was capable of orienting herself in both time and place, speech defects had disappeared, the extrapyramidal rigidity showed some improvement; the sphincters were functioning normally. One year after the start of the treatment she had gained 2 kg. The trophicity of the skin was improved; she now slept eight hours a night.

By *May, 1953* the old woman had become lively, optimistic, and had begun to work within the limits of her vision. The muscular contraction, measured at 10/9 before the stroke, had improved to 18/12. The murmur over the aorta was considerably weaker, the blood vessels less hardened, the hair completely repigmented. Vaginal smear: 15 per cent acidophils, deep cells not present.

*Present condition (1956):*
Muscle power and will to work increased, dynamometry now 20/18.5. Her mood is one of optimism. Blood vessels no longer hardened, active and passive motion in all joints within the physiological limits, weight gain 2 kg. Trophicity of skin improved. Blood pressure 160/90, vision unchanged.

## Case No. 3

T.J., hospitalized, 66-year-old man, admitted June 21, 1952.

*Status praesens:*

Aphasia and amnesia (post-psychic trauma). The patient has great difficulties in speaking, his weight loss is obvious. In view of his mental state he cannot dress himself, cannot solve complicated problems, often loses his continuity of thought, talks to himself, has no power of orientation. The symptoms have developed gradually. At first, speech defects developed. The patient has forgotten proper names and common words. Insomnia. Trembling of arms. Emotional memories make him cry. Five months before admission to the hospital—treatment with Vitamin $B_1$ and male hormones without success. Clinical examination: Sad, frightened look, loss of hair, partial graying, tendency to loss of weight. Extrapyramidal rigidity, particularly on the right side, lack of orientation; he cannot find his room, bent-over body, general weakness in reflexes.

*Diagnosis:*

Pick's syndrome. Mental involution. Laboratory examinations before treatment: Bordet-Wassermann (blood-spinal cord) negative: Nonne-Appelt negative; erythrocyte count 3,090,000, WBC 6600; blood sedimentation 13/23; urea 0.43 per cent; uric acid 0.44 per cent; cholesterol 1.25 $^o/_{oo}$, glycemia 0.95 per cent; Takata-Ara negative; Gros 1.69 and 2.30; total proteins 8.50; dynamometry 21/22; basal metabolic rate +23 per cent; oscillometry: vascular hypotension; lack of vascular reaction in great exertion; vital capacity of 3100, after exertion, 3000; blood pressure 160/70; pulse 62.

In June, 1952, procaine treatments were started. Ten series of twelve injections each were given.

*Result:*

Improved sensitivity, no more depression. Able to orient himself in time and place. Restful sleep, sense of well-being. Able to converse, absorbs new facts. Gain in weight 4.5 kg. Analysis: cholesterol 1.54 $^0/_{00}$, protein 9; Takata-Ara negative; Gros 0.5 and 1.6; basal metabolic rate $+4.2$ per cent; vital capacity 3300, after exertion 4200; dynamometry 25/25. The heart adjusts itself relatively fast to exertions. The same treatment is being continued at home. Physiological and mental state constantly improving. Walks by himself and participates in the family life.

## Case No. 4

T.M., Forty-two-year-old man.

*Diagnosis:*

Premature aging, chronic polyarthritis, polyarthropathy, psoriasis. The history reveals a neuro-arthropathic past: sciatica at 17 years, acute polyarthropathy at 31, clinical and radiological signs of lumbar spondylosis.

Two months after the first symptoms of spondylosis appeared, the skin of the ears broke out, and a year later psoriatic patches could be seen on the outside of the knee and elbow. At the same time the patient complained of joint pains and there were ankylotic processes in the joints (intermetatarsal, tarso-metatarsal, and metacarpocarpal joints). Sedimentation rate was 130 mm in the first hour. Gold and X-ray treatments at the dermatological clinic in 1954 were without results, as were aurothioglucose, sodium salicylate, and multi-vitamin treatments. The patches spread and soon covered almost the whole body, including the scalp. Lesions appeared on the nails; the knee joints were affected, so that the patient could move only with the help of crutches. The procaine treatment was then started.

*Progress:*

After 24 injections the patient could get up by himself, walk without crutches, his joint symptoms were improved, the psoriasis patches were disappearing from the hands. Temporarily the procaine treatment was interrupted by treatment with 800 mg ACTH.

On *July 12, 1955,* a new rheumatic attack occurred; the patient was now treated with procaine exclusively. As a result of treatment the pain and inflammation have disappeared, the patient has gained 6 kg, his sex drive, which was quiescent during the illness, has returned, psoriasis is receding more and more, scalp hair and nails have become normal; blood sedimentation 60/90.

*Present condition:*

Patient is again in a good mood, climbs stairs without difficulty, is able to work again. X-rays reveal that the trabeculae have become denser, the bone structure has become more normal, remineralization of the tibial spines has occurred in the right knee and in the upper trochanteral region. The trophic appearance now is excellent, all signs of premature aging have disappeared, and the patient appears much younger than his real age.

## Case No. 5

P.M., Our oldest patient, has been under treatment since February, 1954. He was born in 1847, and was first seen by us in February, 1954. At that time his body weight was 47 kg, and his diagnosis was striopallidary syndrome, old age.

Because of general debility he was unable to leave his room, suffered from lapses of memory, strong trembling of his hands, characteristic atrophy and scaliness of the skin of his hands, brown spots on the face and on the hands, total achromotrichia, alopecia of the back of the neck, muscular atrophy, particularly of the muscles along the temples, diminished adipose tissues, muscular strength was 14 kg on the right and 12 on the left.

*Cardiovascular system:* Systolic murmur at the focus of the aorta, pulse: 76, blood pressure: 135/70 mm Hg. Arteries along the temples were visibly tortuous, as were those in the shoulder area, with some hardening of arteries in the region of the upper arm.

*Nervous system:* Static trembling of the upper limbs, occasionally involving the head, largely involuntary in nature. General extrapyramidal hypertonia; difficult gait with small steps. Lively deep reflexes, absence of superficial reflexes along the abdomen, plantar reflexes present when feet flexed; sensitivity to vibration absent in the lower left leg.

*Laboratory findings: 1954:* protein 8.00 g per cent, albumin 4.10 g per cent, globulin 3.90 g per cent, A/G ratio 1.05, glucose 82 mg per cent, uric acid 0.037 $^0/_{00}$, cholesterol 1.30 $^0/_{00}$, serum calcium 8.9 mg per cent. Blood analysis: red blood corpuscles 4,200,000, leucocytes 4800, neutrophils 68 per cent, eosinophils 2 per cent, lymphocytes 25 per cent, monocytes 5 per cent, sedimentation rate 10 in 1 hr, 21 in 2 hours, 63 in 24 hours.

Treatment was begun in December, 1954.

*Progress:*

By *March, 1955* his mental condition had improved, he had become more communicative, his gait was improved.

In *October, 1955* a decrease in the trembling of his hands was observed, and his head no longer trembled. He went to town daily. His skin and features had become more trophic, 60 to 70 per cent of the hair growing in then was pigmented and 20 to 30 per cent of the white hair that was present had again become repigmented.

His facial blotches had disappeared by *January, 1956.* The muscles along his temples had redeveloped. By then he was able to take care of himself, his appetite had been restored, he slept 7 hours nightly. His blood pressure had become 150/90 mm Hg.

*By September, 1956* his humeral and radial veins appeared to have become less hardened, while the humeral ones appeared to have become less tortuous. The hair on his forearm had increased and become repigmented to the extent of 90 per cent. Much of his pubic hair had become repigmented.

*In July, 1957* his general attitude was good, his extrapyramidal stiffness had disappeared almost completely, his hands trembled only if he was excited or after an effort, he was in a good mood, communicative and was able to stand upright. His gait was still cautious, but only because of disturbances in his vision (double cataract which had been operated on in 1953). He was able to walk up steps with ease.

*Intercurrent diseases:* grippe with bronchopneumonia in 1955; body weight: 57 kg, muscle power was 17 kg on the right, 16 kg on the left. His blood pressure was 145/80 mm Hg, with a pulse of 77.

*Laboratory findings:* Protein level 7.44 g per cent, albumin 4.40 g per cent, globulin 3.04 g per cent, A/G ratio, 1.45; uric acid: 0.036 $^0/_{00}$, cholesterol 1.96 $^0/_{00}$. Blood analysis: rbc count: 4,300,000, color index 1.01, leucocytes 5300; Band neutrophils 1 per cent, segmented neutrophils 59 per cent, eosinophils 2 per cent, lymphocytes 31 per cent, monocytes 7 per cent.

## Case No. 6

S.P.—Retired professor of mathematics, born in 1859. Hospital admission: September 11, 1952.

*Diagnosis:*

Mental senility, arteritis obliterans along the lower limbs. Mother died at age 98 years.

*Examination in 1953 revealed:* Has had symptoms of intermittent claudication for the past 11 years, starting at first after walking for 200 m, later after 100 m. Facial expression fairly rigid, without muscle tonus. Skin along hands and feet full of senile blotches. The feet were cyanotic and cold, with total achormotrichia and alopecia along forehead, temples and on top of the cranium. Muscular atrophy was pronounced, particularly of the interosseal muscles. Extrapyramidal hypertonia. The patient had difficulty in getting out of bed, complained of pain in the lower leg, even when at rest, seemed very depressed and quite uncommunicative. Cardiovascular system: pulse 60, severe systolic murmur in the focus of the aorta, arteriosclerosis, tortuous veins along the arms. Carotid and axillary arteries sclerotic. Neurological findings: gait painful, jerky; difficulty with delicate finger movements. Deep reflexes: achilles reflex diminished in intensity, palmar and chin reflexes can hardly be elicited. Hernia. Following admission was treated between November, 1952 and June, 1953 with placental extract, which appeared without effect on either his claudication or his general mood and frame of mind. In June, 1953 his weight was 66 kg. His hands and feet were cyanotic and cold, he had cramps and paresthesia, the pulse of the dorsal and posterior tibial arteries in the foot could not be felt. Blood pressure: 135/75 mm Hg. Oscillometric index: Lower leg: upper right third 3, upper left third 1; lower right third ⅓, lower left third ¼; right muscular lift 12.5 kg, left muscular lift: 13 kg.

*Laboratory findings:* protein 7.99 g per cent, albumins 4.2 g per cent, globulins 3.79 g per cent, A/G 1.10; cholesterol 1.92 $^0/_{00}$, glucose 67 mg per cent, urea 0.38 $^0/_{00}$, phosphate 0.032 $^0/_{00}$, serum calcium 9.30 mg per cent.

Intra-arterial novocain treatment was begun on April 30, 1953, with 5 cc of a 2 per cent solution given every other day, the injection alternating between the right and left femoral arteries.

*Progress:*

After 12 injections claudication appeared to occur only after the patient had walked 400 m. On *December 30, 1953* his mental attitude seemed to have improved, he was lively, read novels and works dealing with economics, thought about mathematical problems. In *July, 1954* the trophic condition of his skin seemed to have improved, the senility blotches appeared to be receding and the skin no longer was dry. In *October, 1954* examination revealed that posterior tibial arteries seemed patent. At the same time hair growth and pigmentation were seen to occur in areas which till now had been free of hair.

In *July, 1955* the man began daily visits to town and had no signs either of troubles in his legs or of intermittent claudication. His skin seemed trophic, the facial muscles appeared to have regained their tonicity. *In August, 1956* the xanthelasma was observed to have receded along the inner edge of the right eye and to have disappeared along the outer edge of the left eye.

*By 1957* the patient's complaints were restricted to his hernia. His facial expression had become lively, he had started writing poetry, did much reading, participated in social life. His blood pressure was 125/60 mm Hg in 1957. The oscillometric index was 3.5 in the upper right third of the lower leg, and 2.5 in the upper left third; in the lower right third it was 3, and 2 in the lower left third. His muscle lifting power was 19 kg on the right and 17 kg on the left.

*Laboratory findings:* protein 9.04 g per cent, albumins 5.04 g per cent, globulin 4 g per cent, A/G ratio 1.26. Blood glucose value: 83 mg per cent, uric acid: 0.048 $^0/_{00}$, phosphate 0.026 $^0/_{00}$; serum calcium 10.7 mg per cent.

( The patient died in 1958. )

## Case No. 7

G. L.—A retired woman physician, 63 years old, who was admitted on June 30, 1953.

We have previously reported on this case in collaboration with Parhon, David and Nitea at the National Physicians Congress held in the Spring of 1957 (Gerontological Section), inasmuch as this was an unusual case of parkinsonism and dementia which was cured by two-year treatment with procaine.

Excerpts from the anamnesis: tick typhus at the age of 26 years, but claims no other infectious diseases. Was intensely active in her profession. She was admitted to the hospital of the Institute on June 30, 1953, because of complaints concerning disturbances of her gait, intermittent trembling of hands and lips. Examination revealed a rigid facial expression, bright eyes, the beginning of a cataract in her right eye, masculine hair distribution on the face, uncertain gait with small steps, nightly cramps in her thigh. Blood pressure: 95/55 mm Hg; muscle lifting power 8 kg on the right, 5 kg on the left. Complained of fatigue. Examination of the fundi revealed vascular constrictions.

Following two months of intramuscular treatment with procaine, her condition had improved to the point where she was able to leave the hospital.

She returned to the hospital in August, 1954 with the following complaints: headache, uncertain gait, quick fatigue incurred after even the most modest physical or mental exertion, confusion, tendency to want to sleep.

*Objective findings:* Abasia, astasia, the patient could neither stand nor walk unaided, behavior disturbances (impulsive, euphoric, alternately depressed), more acute reflexes, extra-pyramidal right-handed stiffness.

Procaine treatment was again initiated in doses of 5 cc of a 2 per cent solution given in series of 12 injections, with 10-day rest periods in between.

*Progress:*

The clinical picture that has just been described remained unchanged for the period August, 1954 through July, 1955; her gatism was further accompanied by sphincter incontinence, with total inactivity and lack of memory with respect to fixed associations and arithmetic. She was unable to read or to sign her name. Treatment was continued.

Between July, 1955 and April, 1956 her cerebral function improved gradually; occasionally she showed good capacity for orientation in both space and time. Her ability to do arithmatic and to recognize people returned. She had been unable to leave her bed since admission. When she had at-tempted to do so, an orthostatic syndrome developed. When medication was supplemented with ascorbic acid, the ortho-static syndrome began to disappear and she was again able to walk. Administration of Vitamin C was discontinued, but treatment with procaine was continued. The patient had begun to read and write, and made abstracts of medical articles.

After two years of treatment, all of the symptoms that necessitated her admission into the hospital had disappeared. Her dynamometric achievement increased from 8 to 28 kg on the right side, and from 5 to 20 kg on the left side. Her body weight increased from 48 to 52 kg, and she retained this weight and state of health.

## Case No. 8

M.A.—This 6½-year-old girl was brought to our Institute after various methods of treatment, both drugs and physical methods, had failed over a span of several years.

*Diagnosis:*

[In consultation with Drs. Voiculescu (neurologist) and Constantinescu (pediatrician)]: Hereditary muscular atonia (Oppenheim's), ichthyosis syndrome with keratosis pilaris.

*Excerpts of the previous clinical history:*

Born with a retroversion of the foot which was corrected after 4 months. After the first year of life her head was observed to droop, after the second year, her gait was observed to be uncertain and swaying; she was seen to fall easily and be unable to get up; at the end of the third year, tendinous retraction of the limbs was observed, leading to clubfoot formation and preventing any walking. Family history: Mother was obese, sustained a grippe-like influenza during her third month of pregnancy.

*Clinical examination:*

The child looks her age; height, measured lying down, is 109 cm, standing—104.5 cm. Her body weight 15.6 kg. She cannot stand upright unaided, her swaying gait can be sustained only with help. Her face is pale, the skin rough, and scales easily at the elbow; she can bend her elbow 130°, her fingers 165°. She can bend her thigh at the hip joint up to 130°, but her foot remains in the clubfoot position. Her deep tendon reflexes are weak, her plantar and superficial reflexes cannot be evoked. The Babinski reflex is weakly positive.

Otherwise her mental age corresponds to her chronological one; visceral and neurological examinations were negative. Her blood picture normal, except for a 60 per cent lymphocytosis. X-ray examination reveals spinal kyphosis with minor demineralization. Maximum extension of the right knee 145°, flexion of the right foot 168°.

This was her condition when treatment was begun with 2 per cent procaine solution, given in doses of 2 to 3 cc daily, three times weekly intramuscularly, three times weekly peritendinously. At the same time massage, hot paraffin packs, and physical therapy were begun. It should be added that these latter measures had been taken previously, without accompanying procaine therapy, but had been without success.

*Progress:*

After 60 procaine injections the child has regained liveliness, her skin is again trophic, she can stand upright for short periods, can take a few swaying steps if aided, and is capable of putting her entire left foot down on the floor. Her motion very much improved, she has gained 4.5 cm in height and 2 kg in weight, her muscle tonus has improved, the tendons are less tense, the kyphosis is less marked, she can extend her knee by 8° more, her fingers by 15° more, her thigh by 25° more, can flex her feet up to 140°, and her patellar reflexes have become clear. Her lymphocytosis has receded to 48 per cent, the kyphosis is no longer evident upon roentgenological examination of the spine, the thighs can be extended by 18° more, the foot flexion has increased by 20°.

Thus after 5 years the first signs of improvement can be seen, an improvement that speaks well for the future.

## Case No. 9

V.W.—Paperhanger, 59 years old. Admitted to the hospital on March 30, 1953 because of complaints of loss of weight, of swelling and loss of function in arms, of stiff gait, of trembling of hands. The patient had been treated for acute hyperthyreosis during the preceding three months, the joint condition having developed suddenly in March, 1953.

*Clinical examination* revealed premature aging, rigid facial expression, glassy eyes, general hypotrophy, a very aged appearance of face and skin, with greying of the eyes. The joints of the arms and legs were painful and swollen, and motion was limited—particularly in the hands, which showed deformities. The pulse was 120/min., there was hardening of the major arteries. The sedimentation rate was 25/48, the BMR was 40 per cent, blood protein 7.81 per cent, albumin/globulin ratio 1.06. Blood picture was normal with 43 per cent lymphocytes. X-ray findings: trabecular structure unclear, cortical bone thinned appreciably. Bones had a glassy appearance.

*Progress:*

After five months of antithyreotoxic treatment (Lugol's solution, Alkyron, sedatives), his thyroid symptoms receded somewhat, but his joint symptoms became more severe. Procaine treatment was therefore initiated and the antithyreotoxic treatment was reduced. After three series (36 procaine injections) the improvement was appreciable: his trophicity had improved (there was an 18 kg increase in weight), his skin had become more elastic, his senile appearance had disappeared, his joint symptoms had receded. A very slight stiffening of the small joints remained, but did not prevent him from working at his trade.

*Laboratory findings* were normal: sedimentation rate: 6/24 mm, lymphocytes 27 per cent, blood protein 8.05 per cent, A/G ratio 1.65. X-ray examination revealed a thickening of the trabeculae and remineralization.

## Case No. 10

J.E.—Seventy-year-old housewife, admitted to the hospital in 1949 because of two years of complaints about the right knee, with difficulty in walking. Examination revealed signs of aging, particularly of the skin and face, and atrophy of the subcutaneous connective tissue and muscles, along with appreciable and painful swelling of the right knee joint with fixation in a semiflexed position. Examination also revealed hardened, tortuous arteries that could be felt, and slight signs of extrapyramidal rigidity.

*Sedimentation rate:* 41 mm in the first hour; total protein 7.99 g per cent, of which 4.42 g per cent was albumin and 3.57 g per cent was globulin. A/G ratio 1.23. Blood cholesterol 144 mg per cent.

Following 4 intra-arterial injections of a 1 per cent procaine solution, the pain in the knee had stopped, motion was entirely restored, the swelling had receded and the patient was able to walk again.

Intramuscular procaine injections were continued thereafter at irregular intervals. Slowly the trophicity of the skin improved, and the patient gained 4 kg weight. During a relatively long interruption in treatment the patient developed fever on November 29, 1951, followed by light but painful swelling of the right knee joint. Following three intra-arterial injections these symptoms disappeared completely. The patient subsequently maintained her treatment schedule and presented an entirely different picture by July, 1955: She had gained 8 kg, her skin had become elastic, color had returned to her hair, her capacity to work had increased. Her mental capacity had improved (return of memory), she had better muscle strength, and her voice had become higher. The joint pains had disappeared completely. Sedimentation rate: 25 mm in the first hour, protein level 7.59 g per cent, albumins 4.43 g per cent, globulins: 3.16 g per cent, A/G ratio 1.40: cholesterol 187 mg per cent. Her condition is the same at this writing.

## Case History No. 11

*(From the files of the Leningrad Sanitary-Hygiene Medical Institute, Leningrad, USSR, as reported by N. K. Gorbadei.)*

N.N.V.—Male, aged 38 years. Admitted to the therapeutic clinic of the Leningrad Sanitary-Hygiene Medical Institute on August 16, 1954 with a diagnosis of exacerbation of toxic-infective polyarthritis.

He came to hospital on crutches, being unable to walk. He complained of severe pains in the knees, the left ankle, elbow, wrist and finger joints and in the right shoulder. All these joints were swollen and painful. The patient was unable to extend his feet or his knees. His temperature every day throughout the months was between 100.4° and 102.2°. Being confined to his room for treatment for this month, every day the patient received from 6 to 12 g of sodium salicylate, but with no effect. The ESR on admission was 64 mm in 1 hour and the white cell count 10,200. Roentgenograms of the right and left knee joints showed no destruction of bone. At the upper and lower poles of the patella there were spicules of bone.

The pains in the joints first appeared in December, 1950. The onset of the disease was associated with frequent attacks of influenza and tonsillitis. From December, 1950 to March, 1951 he was in the Postgraduate Medical Institute. He had an exacerbation of the condition in October-November, 1951. After this he felt the pains in all the joints only from time to time, and he continued to work. Once he had a relapse after bathing in the river during June, 1954.

Despite the fact that the patient had previously been given large doses of salicylates, ultrasonics, intramuscular penicillin, erythema doses of ultraviolet light, his condition remained poor. The pain and swelling in the joints persisted, and his temperature was between 99.9° and 102.2° every day. He never got up, even on crutches, because of the severe pain and contractures of the knee joints. In view of the lack of response to treatment, on September 17, 1954, i. e., after one month, it was decided to give the patient a course of intra-arterial infusions of penicillin with procaine.

After he had received the first 7-8 infusions of procaine with penicillin, the pains in the patient's joints were greatly diminished and his temperature fell to normal. The patient began to walk, at first with crutches and later with a cane. Altogether he received 20 intra-arterial infusions of penicillin with procaine, each of 200,000 to 800,000 i.u. of penicillin.

As a result of the treatment, the pain and swelling of the patient's joints completely disappeared. The movements of the joints became free and painless. The patient began to walk unaided along the corridor and in the ward. His temperature remained persistently normal. The ESR fell from 64 mm to 28 mm in 1 hour. From a leucocytosis of 10,200, the white cell count fell to 4500.

On November 6, 1954 the patient went home unaided, without a cane, in a satisfactory condition. Five months later, when the patient was recalled to hospital, he reported that the state of his health was good. According to present information, the patient remains well and is continuing to work.

## Case History No. 12

*(From the files of the Leningrad Sanitary-Hygiene Medical Institute, Leningrad, USSR, as reported by N. K. Gorbadei.)*

K., male, aged 34 years. Admitted to the therapeutic clinic of the hospital on December 6, 1953 with the diagnosis of bronchial asthma, emphysema of the lungs, deformity of the chest following wounding, atherosclerosis and cardiosclerosis.

On admission, he complained of attacks of severe breathlessness, as many as 6 times a day, constant shortness of breath at times other than during an attack, cough with a large amount of mucopurulent sputum, and stabbing pains in the left half of the chest.

He had felt unwell since 1950. He associated the onset of his illness with compression and deformity of the chest in a tank in wartime, which was followed by bilateral pneumonia. When in the army, he also had a severe contusion. He served in the army from 1939 to 1949.

He associated the latest exacerbation with an attack of influenza. Because of his attacks he frequently summoned the emergency aid service, sometimes as often as 3 times a day.

On admission, the patient's condition was serious. He did not remain free from attacks for long. His shortness of breath was acute, and his lips, nose and fingers were cyanosed. Marked deformity of the thoracic cage; sternum pressed inward. Box-like percussion sounds over the lungs, on auscultation—against a background of harsh breathing, many scattered dry sibilant rales could be heard. The mobility of the borders of the lungs was severely restricted. On account of emphysema the borders of the heart could not be accurately defined. The heart sounds were muffled. Pulse 98-100 per minute, regular, volume and tension adequate. Abdomen soft and not tender. Liver at the costal margin. No edema of the legs.

*Fluoroscopy:*

Anteriorly, on the right side, a small area of thickened pleura in the hilar zone and widening of the lower pole of the right hilum were seen. The lungs were emphysematous. Movement of the diaphragm was limited, more so on the right. No Charcot-Leyden crystals nor Curschmann's spirals were found in the sputum. On admission the ESR was 18 mm in 1 hour. Eosinophils—2 per cent.

At first the patient was given oxygen, intramuscular penicillin and subcutaneous injections of camphor and adrenalin; he inhaled the smoke from asthma powders, but the attacks persisted and there was no decrease in the shortness of breath.

On December 15, 1953 the patient began a course of intra-arterial infusions of 0.5 per cent procaine solution, the dose being 30 cc each infusion. After the first 6-7 infusions the patient felt considerable relief of his dyspnea and he slept better. Whereas before treatment he woke seven times during the night to smoke his asthma powder, after treatment he slept the night through.

Two weeks later, i. e., on December 30, 1953, the patient's attacks of breathlessness ceased quite suddenly, despite the fact that, in addition to the intra-arterial infusions of procaine, he was receiving no other form of treatment. During his stay in hospital the patient gained 13 pounds in weight.

Altogether the patient received 13 intra-arterial infusions of 25 to 30 cc of 0.5 per cent procaine solution. One month after the start of treatment the patient was discharged from the clinic in good condition, with no attacks of breathlessness.

Three years have now elapsed since this patient's discharge from the clinic, and he still feels well and is free from attacks of breathlessness and dyspnea.

# Bibliography

# Bibliography

ALTHOFF, W.: Die therapeutische Novocain-Anwendung in der inneren Medizin. (Darmstadt, 1947).

ASLAN, ANNA: Eine neue Methode zur Prophylaxe und Behandlung des Alters mit Novocain—Stoff H3—eutrophische und verjüngende Wirkung. *Die Therapiewoche*, Vol. VII, No. 1/2 (1956).*

ASLAN, ANNA: Novocain als eutrophischer Faktor und die Möglichkeit einer Verlängerung der Lebensdauer. *Therapeutische Umschau*, Vol. XIII, No. 9 (1956).

ASLAN, ANNA: Geriatrie in Rumänien. *Medizinische Klinik*, Vol. 52, No. 40 (1957).

ASLAN, ANNA: Recherches sur l'action de la novocaine—action eutrophic et rejeunissante. Report to 4th Gerontological Congress in Merano, Italy (1957).

ASLAN, ANNA: Neue Erfahrungen über die verjüngende Wirkung des Novocains—Stoff H3—nebst experimentellen klinischen und statistischen Hinweisen. *Die Therapiewoche*, Vol. VIII, No. 1 (1957).*

ASLAN, ANNA: Behandlung der Alterserscheinungen mit Novocain. *Der Praktische Arzt*, No. 126 (November, 1957).

ASLAN, ANNA: Zur Wirkung des Novocains. *Arzneimittel-Forschung*, Vol. VIII, No. 1 (1958).

ASLAN, ANNA: Die Wirkung von Novocain und PAB auf den Sauerstoffverbrauch der Bierhefe. *Arzneimittel-Forschung*, Vol. VIII, No. 3 (1958).

ASLAN, ANNA: La novocaine—substance $H_3$—dans la thérapeutice de la vieillesse. *Revue Francaise de Gérontologie*, Vol. IV, No. 10 (1958).

ASLAN, ANNA: Richerche sull'azione della novocaina. Azione eutrofica e di ringiovanimento. *Minerva*, Vol. XLIX, No. 75 (1958).

197

ASLAN, ANNA AND S. CAMPEANU: Die Wirkung von Novocain und p-Amino Benzosäure auf den Sauerstoffverbrauch der Bierhefe. *Arzneimittel-Forschung*, Vol. VIII, No. 8 (1958).

ASLAN, ANNA AND C. DAVID: Le traitement de l'ictus par la pro-caine—substance $H_3$. *Revue Francaise de Gérontologie*, Special Issue (1959).

ASLAN, ANNA, C. DAVID AND S. CAMPEANU: Der Einfluss der Dauer-behandlung mit Novocain auf die differenzierte Kapillarper-meabilität. *Arzneimittel-Forschung*, Vol. IX, No. 9 (1959).

ASLAN, ANNA, C. DAVID AND S. CAMPEANU: Die Verschiebungen der Kapillarpermeabilität bei Arteriosklerose. *Arzneimittel-Forschung*, Vol. IX, No. 9 (1959).

BRAUNSTEINER, P.: Gerioptil bei der Behandlung der Altersschwer-hörigkeit. *Die Medizinische*, No. 36 (1958).

*British Medical Journal* (Editorial), November 28, 1959, p. 1163.

*British Medical Journal* (Special Report), November 28, 1959, p. 1175.

DAVID, CORNEL: Biochemische und physiologische Alterskriterien und ihre Veränderungen im Laufe der Behandlung, als Rehabi-litationsmerkmale, aus der Erfahrung des Instituts für Geriatrie. *Die Therapiewoche*, Vol. VIII, No. 1 (1957).*

EICHHOLTZ, F.: Die Anwendung von Novocain in der inneren Medizin. *Klinische Wochenschrift*, Vol. XXVIII, No. 45/46 (1950); Vol. XXX, No. 5/6 (1952).

FENZ, E.: Behandlung rheumatischer Erkrankungen durch Anä-sthesie. (Leipzig, 1941.)

GASSMANN, M., R. JAQUEROD, O. LAEPPLE AND R. SCHAEFER: *Schweizerische Medizinische Wochenschrift*, Vol. 88, No. 29, p. 716 (1959).

GOOD, M. G.: Das Problem des Rheumatismuses. *Deutsche Medi-zinische Wochenschrift*, Vol. XXV, No. 25 (1951).

GOHRBANDT, E.: Zur Behandlung der multiplen Sklerose. *Zentral-blatt der Chirurgie*, p. 726 (1950).

GORBADEI, N. K.: Intraarterial Infusion of Procaine in Therapeutic Practice. (Translated from the Russian—Consultants Bureau, Inc., New York, 1960.)

GRAUBARD, D. J. AND M. C. PETERSON: Clinical Uses of Intravenous Procaine. (Thomas, Springfield, Ill., 1950.)

GREENE, R., J. VAUGHAN-MORGAN AND J. GAMMON: The Effect on Bleeding Time of an Extract of Blood of Patients with Rheumatoid Arthritis. *British Medical Journal*, January 5, 1952, p. 17.

HUNEKE, WALTER: Verjüngung durch Novocain. Bestätigung, Kritik und Ausblick auf weitere Forschungsaufgaben. *Hippokrates*, Vol. XXIX, No. 4 (1958).

HUNEKE, WALTER: Verjüngung durch Novocain (Stuttgart, 1959).

KOHLER, UDO AND E. MAMPEL: Erste Erfahrungen mit der Novocainbehandlung alter Menschen. *Die Therapiewoche*, Vol. VIII, No. 1 (1957).

KOHLER, UDO: Verjüngung mit Stoff $H_3$"? *Der Landarzt*, Vol. XXXIII, No. 33 (1957).

KOHLER, UDO: Die moderne Behandlung von Altersleiden mit "Stoff $H_3$". *Hippokrates*, Vol. XXIX, No. 1 (1958).

KOHLER, UDO: Neue therapeutische Möglichkeiten bei Alters- und Abnutzungsleiden. *Zeitschrift f. Arztliche Fortbildung*, No. 2 (1958).

KOHLER, UDO: Perorale Therapie der Alters- und Abnutzungsleiden. *Arztliche Praxis*, Vol. X, No. 52 (1958).

KOHLER, UDO: Geriatrie in Rumänien. *Das Deutsche Gesundheitswesen*, Vol. 14, No. 4 (1959).

*Lancet* (Editorial), March 14, 1959.

LEAKE, CHAUNCEY: Russian and Iron Curtain Proposals for Geriatric Therapy. *Geriatrics*, Vol. 14, No. 10 (1959).

LILJESTRAND, G. AND R. MAGNUS: *Münchner Medizinische Wochenschrift*, p. 551 (1919).

LUTH, PAUL: Möglichkeiten und Aussichten der Altersbehandlung nach Parhon und Aslan. *Deutsches Medizinisches Journal*, No. 8 (1958).

LUTH, PAUL: Gerotherapeutische Erfahrungen mit "Gerioptil pro injectione". *Medizinische Klinik*, Vol. LIII, No. 29 (1958).

MAINLAND, DONALD: Elementary Medical Statistics, W. B. Saunders Co., Philadelphia, 1952.

PARHON, C. I.: *Die Therapiewoche*, Vol. 8, No. 11 (1957).*

---

*Appears in English translation "Research on Novocain Therapy in Old Age" (Consultants Bureau, Inc., New York.)*

PETERSEN, F.: Novocain-Behandlung alter Menschen in der ner-
venärztlichen Praxis. *Medizinische Monatsschrift*, Vol. XIII,
No. 5 (1959).

PFEIFFER, CARL L. AND ASSOCIATES: Stimulant Effect of 2-Dimethyl-
aminoethanol — Possible Precursor of Brain Acetylcholine.
*Science*, Vol. 126, No. 3274 (1957).

PORTIAS, H.: Sur une expérience de 86 cas de traitements de
troubles de la sénescence par la procaine. *Revue Francaise de
Gérontolgie*, Vol. V, No. 4 (1959).

SAMSON, E. I.: Changes in the Motor and Secretory Gastric Functions Following Intravenous Injection of Novocain. *Pharmacology and Toxicology*, No. 3 (1959).

SOHRING, K.: Novocain, seine Wirkung und Anwendung. *Pharmazie*,
Vol. IV, Nos. 7, 8, 9 (1949).

UDALOV, YU. F.: Effect of Novocain on Tolerance of White Rats to
High Altitudes. *Bulletin of Experimental Biology and Medicine*,
Vol. XLI, No. 8 (1956).

UDINTSEV, G. N. AND V. B. BLANK. Cited by Gorbadei, N. K.

UNGAR, GEORGES: Experimental and Traumatic "Shock". Factors
Affecting Mortality and Effect of Therapeutic Agents (Ascorbic
Acid and Nupercaine). *Lancet*, Vol. 1 (1943) p. 421.

UNGAR, GEORGES: The Inhibition of Histamine Release by a Pituitary Adrenal Mechanism. *Journal of Physiology*, Vol. 103, No.
3 (1944) p. 333.

# Glossary of medical terms

# Glossary of medical terms

**adrenals:** Two glands located over each kidney, secreting adrenalin and other hormones.

**adrenocortical hormones:** Secretions from the pituitary gland acting on the adrenal glands, the best known of which is ACTH.

**albumin:** Protein substance composing the largest part of the tissues.

**amenorrhea:** Absence of menstruation before the advent of menopause.

**analgesic:** Possessing pain killing property.

**anaphylactic shock:** An attack, sometimes fatal, due to injection of a substance to which the body had become sensitized through an earlier injection.

**angina pectoris:** Attacks of severe, constricting pains in the chest, caused by lack of oxygen in the heart muscle.

**ankylosing spondylitis:** Arthritis of the spine, accompanied by ossification of ligaments and stiffness of neck.

**antiphlogistic:** Acting against inflammations.

**arrhythmia:** Variation from normal rhythm of the heart beat.

**bursitis:** Inflammation of sac (bursa), usually between joints and filled with fluid to reduce friction.

**capillaries:** Small blood vessels connecting arteries and veins.

203

**central nervous system:** One of the two main divisions of the nervous system, consisting of brain and spinal cord.

**cerebral cortex:** Outer layer of the brain.

**claudication:** Limping.

**clavicle:** Collarbone.

**cornea:** Transparent outer layer on the front of the eyeball.

**costal cartilages:** Elastic tissue forming the ribs.

**cubital vein:** The elbow vein.

**cyanotic:** Skin and mucuous membranes being bluish, due to lack of oxygen in the blood.

**diencephalon:** The interbrain.

**dysmenorrhea:** Difficult, often painful menstruation.

**dystrophy:** Degeneration of any part of the body due to faulty nutrition.

**endocrine glands:** Glands which secrete their hormones directly into the bloodstream, also called ductless glands.

**endocrinologist:** Specialist in the science of endocrine glands.

**estrogens:** Hormones produced by the ovaries.

**etiology:** Study of the causes of disease; also causes of disease.

**euphoria:** State of elation.

**eutrophic:** Being in a condition of healthy nutrition.

**extra-pyramidal disturbances:** Syndrome manifesting itself in stiffness of muscles, lack of mobility, masklike facial expression.

**ganglion:** Mass of nerve tissue receiving and emitting nerve impulses.

**ganglion stellatum:** Nerve tissue in the sympathetic trunk, between the upper and lower cervical (neck) ganglion.

**gatism:** Rectal incontinence.

**geriatrics:** That branch of medicine concerned with old age and its diseases.

**gerontology:** Study of the phenomenon of aging.

**globulins:** A group of proteins not soluble in water.

**granulocytes:** White blood cells containing granules, formed in the bone marrow.

**herpes zoster:** Virus disease of the skin, manifesting itself in large sores.

**hyperesthesia:** Generally unusual sensitivity to physiological sensations; here, to drugs.

**ichthyosis:** Fish scale disease, characterized by keratinization of the skin, appearing at an early age and usually lasting through life.

**intercostal neuralgia:** Neuralgia of nerves passing through the ribs.

**intestinal flora:** Totality of bacteria in the intestine.

**keratinization:** The skin becoming horny.

**keratosis:** Any skin disease characterized by overgrowth of horny cells.

**Korsakov syndrome:** Chronic alcoholic delirium.

**labia minora:** One of the two skin folds at the inner surface of the external female genital organ.

**leucocytes:** White blood corpuscles.

**luxation:** Dislocation of bones which form a joint.

**Ménière's syndrome:** Illness manifesting itself in dizziness, nausea, vomiting and partial deafness, usually occurring in advanced age.

**metabolism:** The physiological process of providing energy out of foodstuffs and removing waste products from the body.

**monocytes:** Large leucocytes, making up about 5 per cent of normal blood.

**muscular atony:** Lack of muscle tone.

**myocardial:** Affecting the heart muscle.

**necrosis:** Death of body tissue.

**occiput:** Back part of the skull.

**osteoporosis:** A degenerative disease of the bones, caused by loss of calcium, quite frequent in old age.

**peri-articular:** Around a joint.

**pigmentation:** Existence, or appearance, of coloring matter in cells and tissues (skin, hair, nerves).

**pineal body:** Small gland in the back of the brain with unknown function.

**polyarthritis:** An arthritic condition affecting several joints.

**pruritus:** Strong itch.

**psoriasis:** Chronic skin disease, characterized by reddish patches with white scales spreading all over the body.

**sebacious glands:** Oil producing glands on the skin.

**serum disease:** An allergic reaction due to injection of a serum, manifesting itself in fever, hives and, sometimes, enlarged glands.

**spasmolytic:** Relaxing a spasm.

**sternum:** Breast bone to which the ribs attach.

**striated muscle:** Muscles which are voluntary; under the microscope they have a striped appearance.

**subcutaneous:** Under the skin.

**substitution therapy:** A permanent treatment, substituting medication for a mal- or nonfunctioning organ or gland.

**thrombophlebitis:** Inflammation of a vein, caused by a clot.

**thyroid gland:** An endocrine gland in front of the windpipe.

**trigeminal neuralgia:** Neuralgia of the fifth cranial nerve which serves the forehead, cheeks, teeth and tongue—one of the most painful of all diseases.

**vasoligation:** Ligation of the vas deferens, the ducts which carry the sperms from the testes to the seminal vesicles.